To Jan and Bill Shupert —

May all your travels be safe!

LINCOLN HIGHWAY

The Road My Father Traveled

Jewell

WHAT IS IT ABOUT THE WEST THAT HAS INSPIRED SIREN CALLS to homo sapiens from time immemorial? Marco Polo notwithstanding, the predominant draw has been to the West. Even religious characters like Abraham; father of Judaism, Christianity, and Islam, was not immune to the western magnet, and left Ur of the Chaldees to go west. According to biblical New Testament accounts, the new Judeo/Christian faith went west to Antioch, Rome, and eventually Europe and North America. The adventurous Norwegians sailed west to Newfoundland, and some historians claim they were the first to North America long before Spain's Christopher Columbus went west. People from "western" Europe went west to the "New World" in North America. Once in eastern North America, drawn like lemmings by some undefinable instinct, they went west with little excuse and often without one.

Of course the discovery of silver in Nevada and gold in California didn't hurt. The puffing of railroads made the compulsion more rational, as did the later invention of the automobile. Even more rational yet was the designation of the Lincoln Highway (1913) as the first well promoted coast-to-coast highway. Later President Eisenhower threw all rationality to the winds (some would say) when he genuflexed to Detroit and christened the interstate highway system making it even easier to go west. Perhaps people go west for the same reason people climb a mountain—because it's there.

In 1915, my father was not immune to the western magnet either. His excuse was the Panama-Pacific International Exposition in San Francisco, honoring the completion of the Panama Canal. Obviously, the Exposition was an excuse. Why else would he carve out nearly a year of his life at age thirty-three, endure extreme cold in Wyoming, impossible sand, mud, and salt flats in Nebraska and Nevada, put up with broken springs, clutches, insufficient gas, loneliness, invasive dust, and serious risk to his life by, not only a hostile desert environment, but highwaymen and possibly even hostile Indians (though remote by 1915), to spend only one day at the exposition? It was a rational excuse to go west as millions did before him, and as millions are still doing. My hat's off to him. I'm glad he did.

LINCOLN HIGHWAY

The Road My Father Traveled

D. LOWELL NISSLEY

Copyright © 2006 D. Lowell Nissley

All rights reserved. No part of this book may be reproduced or transmitted in any form or by any means, electronic or mechanical, including photocopying, recording, or by any information storage and retrieval system, without permission in writing from the publisher.

Published by Robert B. Abel
811 Richardson Way
Sarasota, FL 34232

Publisher's Cataloging-in-Publication Data
Nissley, D. Lowell (David Lowell), 1921- .

Lincoln Highway : the road my father traveled / D. Lowell Nissley. – Sarasota, FL : Robert B. Abel, 2006.

p. ; cm.
Includes bibliographical references and index.
ISBN: 0-9662719-5-5
ISBN13: 978-0-9662719-5-9

1. Lincoln Highway—History. 2. United States—Description and travel. 3. Automobile travel—United States—History—20th century. I. Title.

HE356.L7 N57 2006
388.10973—dc22   2006927538

Printed in the United States of America
10 09 08 07 06 • 5 4 3 2 1

CONTENTS

Preface—*ii*

Acknowledgements—*viii*

Introduction—1

PENNSYLVANIA
and the Lincoln Highway—7
Concrete posts of Pennsylvania—11
Does The Old Geezer Still Drive?—15
Columbia/Wrightsville—17
Gettysburg—24
Chambersburg—30
Everett—33
Bedford—38
Burma Shave—43

WEST VIRGINIA
and the Lincoln Highway—44
Chester—44
Autocar Trucks—46

OHIO
and the Lincoln Highway—47
Concrete posts and pillars of Ohio—48
Lisbon—52
Minerva—54
Wooster—58
Packard—60

INDIANA
and the Lincoln Highway—63
Concrete posts of Indiana—67
Fort Wayne—68
La Porte—70

Pre WW II Auto Mfg.—72

ILLINOIS
and the Lincoln Highway—73
Concrete posts of Illinois—74
Chicago Heights—75
Geneva—77
Fulton—79
Importance of Good Roads—81

IOWA
and the Lincoln Highway—82
Concrete posts of Iowa—85
Cedar Rapids—87
Belle Plaine—91
Scranton—93
Carrol—95
Arion/Dow City—99
Dunlap—103
Missouri Valley—105

1915 Autos manufactured—107

NEBRASKA
and the Lincoln Highway—108
Concrete posts of Nebraska—109
Omaha—114
Valley—117
Schuyler—119
Shelton—122
Kearney—125
Elm Creek—128
Overton—130
Paxton—132
Sidney—136
Dirt Roads—138

WYOMING
and the Lincoln Highway—139
Concrete posts of Wyoming—144
Pine Bluffs—145
Burns—148
Carbon—151
Wamsutter—154
Evanston—156
You Auto Know—159

UTAH
and the Lincoln Highway—160
Salt Lake City—168

NEVADA
and the Lincoln Highway—172
Eureka—176
Austin—179
Fallon—183
My Auto, Without Thee—188

CALIFORNIA
and the Lincoln Highway—189
Loomis—194
Oakland—197
San Francisco—199

Frank C. Nissley—204
Post Script—205
Resources—206
Index—210

All photographs are by the author unless otherwise noted

Dedication

To Miriam, My Wife, The computer widow, patient and long suffering with piles of photos and papers on every flat surface in the house, genuinely enjoying the mutual experiences of interviewing, traveling, scanning library microfiche and learning to know the wonderful people of the Lincoln Highway. This has truly been an extraordinary family adventure.

ACKNOWLEDGMENTS

HAVE YOU EVER CONSIDERED the impossible task of thanking your teachers, acquaintances, family, friends, co-workers and myriads of other people who have made us who we are? Someone has said, we are part of all we meet. It is a no less daunting task to adequately give credit to everyone who has contributed to this volume. Like life, many of those who contributed so much were not even aware they were doing so.

It would have been helpful to have interviewed my father about his 1915 adventure. Alas, I was never curious enough to ask questions when questions should have been asked. Nonetheless in his own way he contributed to the making of history even though at the time he was unaware of it. To him I will ever be grateful. His diary opened a crack through which could be seen the lifestyles, values, ambitions, and even warts of those ordinary and in many cases extraordinary people whose presence strung along a narrow path from the Atlantic Ocean to the Pacific, without whom the Lincoln Highway would have remained an empty vision.

Following in his train are the dozens of other people who either knowingly or unknowingly contributed to this brief, narrowly focused history of the Lincoln Highway. Of the Lincoln Highway purists, Jay Banta at Fish Springs, Utah, planted the first seed when my wife and I first met him on our first Lincoln Highway trip, He was the first person we saw, heading west from Vernon, Utah, the first to fill in the blanks in our knowledge of the Lincoln Highway, and the first to suggest a report to the *Lincoln Highway Forum.*

Other Lincoln Highway experts come to mind like Olga Herbert from the Pennsylvania office of the Lincoln Highway Heritage Corridor. The first time we met she had waited over an hour while we unsuccessfully tried to coordinate travel time from Sarasota, Florida to her office, then in Greensburg. She graciously received us even though she and her husband had tickets for a special event that evening. Also in Pennsylvania was Joe Riggles from the Pioneer Historical Society in Bedford, a repository for Lincoln Highway history and events,

Additional Lincoln experts who gave encouragement and counsel were Peter Youngman, Indiana, Brian Butko, Pennsylvania, Michael Buettner and Bob Lichty, Ohio; Ruth Frantz and Lynn Asp, Illinois; Bob Owens, Iowa; Bob Stubblefield, Nebraska; Chris Plummer, Wyoming; Norman Root, California; Tom Lutzi, Nebraska; and Gregory Franzwa, for most everywhere.

And how would any of us with only a smidgeon of experiential interest in history survive without historical societies and museums? An image of Elwood W. Christ emerges at the Adams County Historical Society, Gettysburg, Pennsylvania; Alice Barnes, Loomis Basin Historical Society, Loomis, California; Jeffery Whetstone, Bloody Run Historical Society, Everett, Pennsylvania; Shirley Walker, Churchill Economic Development Authority and Small Business Development Center, Fallon, Nevada; Joy Snowden, Eureka County Economic

ACKNOWLEDGMENTS

Development Council, Eureka, Nevada; and Nadine Beran, Schuyler Museum, Schuyler, Nebraska; James Winter, of the Central Pacific Railroad Photo History Library; Peter Epstein for photos of Donner Pass snowsheds, lake and tunnels; Oakland History Room, Oakland, California Public Library; Hank Zaletel, Iowa Department of Transportation. The Special Collections Library, University of Michigan; Nebraska State Historical Society, Lincoln, Nebraska; Sarah Wesson, Davenport Public Library, Davenport, Iowa; Department of State Parks and Cultural Resources, Wyoming State Archives, Cheyenne, Wyoming; Cheyenne County Historical Association, Sidney, Nebraska; James Kerwin, Carroll, Iowa; Cindy Simon, Arion/Dow City, Iowa.

And we dare not pass by the chambers of commerce. In particular, H. W. Trapnell of the Greater Austin Chamber of Commerce did the chambers proud by going the proverbial extra mile.

Private citizens who had no ax to grind became personal friends in the gentle probing of intimate community affairs and history. Stan Taggart from Evanston, Wyoming, flashes to mind. He is no stranger to the Lincoln Highway having basked in the honor of growing up in a town named, "Taggart," where his father operated cabins and a filling station in Utah. Merrill G. Sargent made Pine Bluffs, Wyoming, come to life. Fae Christensen in Paxton, Nebraska, became a close friend long before we ever met in person. And how would we ever have become privy to the life of George Meisner in Shelton, Nebraska, if not for Jane Bernhard? It was Mildred Heath, editor of the *Overton Observer* for seventy-five years, who marched down the alley with my wife and me in tow for lunch with some local senior citizens. Helen Casper provided a photo of Elm Creek, Nebraska, as a 1915 bird would have seen it. Bev and Wallace Winkie in Belle Plaine, Iowa, were not outdone by anyone in hospitality and eagerness to help. And the delightful story of the Lincoln Highway Christmas in 1926 west of Cedar Rapids, Iowa, was the contribution of Nova Dannels.

Thanks to the many unnamed local citizens in restaurants, business offices, small museums, parks, and even people on the street. The risk is the omission of significant contributors. For those missed, I'm sorry and will trust in their grace to understand.

Sandy Limont, sister Naomi's husband, brought many years of professional graphic arts to bear on this story of my father and the Lincoln Highway. He is a graduate of the Philadelphia College of Art, and has been a senior art director for major corporations in addition to managing his own graphic design studio. Without his skill, patience and willingness to adjust to my whims this volume would have been left to the mercy of much lesser ilk.

Michael Shenk, another brother-in-law, and his wife, Peggy, were the first to read and edit an early text thus encouraging me in the right direction. I plead guilty to the old maxim, "Everyone has a right to his own stupid opinions", so mistakes and goofs belong not at the feet of in-laws.

PANORAMIC VIEW OF MAIN EXHIBIT PALACES OF THE
PANAMA-PACIFIC INTERNATIONAL EXPOSITION, 1915

Courtesy of the Oakland Public Library, Oakland History Room

ORIGINAL 1915 LINCOLN HIGHWAY ROUTE

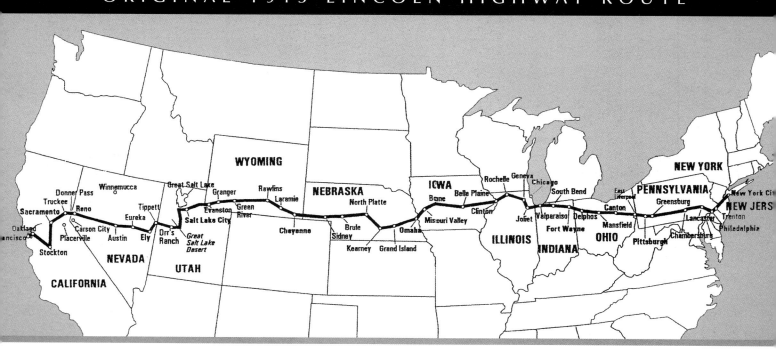

LINCOLN HIGHWAY

The Road My Father Traveled

INTRODUCTION

MANY FINE BOOKS HAVE ALREADY been written about the incomparable Lincoln Highway by highly qualified authors such as Gregory Franzwa, Drake Hokanson, Brian Butko and others. This volume intends to paint Lincoln Highway history with a broad brush rather than repeat minute details. The creation of this book was inspired by my father's diary in 1915, which records his drive in a Little Giant truck from eastern Pennsylvania to San Francisco. The focus here is not on every town on the Lincoln Highway, but on Father's diary and the drama he inadvertently introduced us to in the towns and ranches, many of which no longer exist. There he found shelter, hospitality, and livelihood.

In the late 1800s my father, Frank C. Nissley, was a typical farmboy in Lancaster County, Pennsylvania. At age twenty-one he traded his pitchfork for a camera and became an itinerant photographer, specializing in children, going from door to door photographing them in their homes (or outside if the weather was good). He had "Callers Out" who would make the initial sales contact with the child's mother. If the response was favorable, a chalk mark was made on the sidewalk to signal my father coming along maybe thirty minutes later, allowing time to wash and dress the children. After a few days he would return with proofs for the family to place their order. On some occasions he would rent a local theater and show childrens' pictures on the screen before taking orders.

His gate receipts would be split with the theater. Such shows would often be accompanied by a short humorous film for entertainment, hence the name of his firm, "Novelty Photo & Amusement Company." His diary frequently says, "Showed Victory." "Victory" was no doubt a production featuring Victory Bateman, so named because she was born on the day General Lee surrendered to Ulysses S. Grant. She was an important stage star in the 1890s and made her screen debut in the Thanhouser Company's highly acclaimed two-reel production of Charles Dickens, *Nicholas Nickleby,* in 1912. Over the years our family transported boxes of unlabeled photographs from place to place, only occasionally venturing into them in search of some particular memory. On one such foray I chanced upon a small brown leather diary. It contained notes from my father's trip to San Francisco over the Lincoln Highway in 1915.

On Monday, April 5, 1915, Frank C. Nissley bought a Little Giant truck in Philadelphia, for $1,240, including a 10 percent discount. On June 4, 1915, he pointed his truck westward on the Lincoln Highway and headed for the Panama-Pacific International Exposition in San Francisco, California, where the completion of the Panama Canal was celebrated.

My father's 1915 saga was great, courageous, and timely. It was a response to the call of the pioneer. But on the canvass of the Lincoln Highway his trip was not the only brush stroking the landscape. Although people had been crossing the country via the Lincoln Highway before 1915, that year was a special magnet year since it coincided with the Panama-Pacific International Exposition in San Francisco, drawing many people across the country on the Lincoln in those elegant vintage "machines," the very latest and best autos magicians could produce.

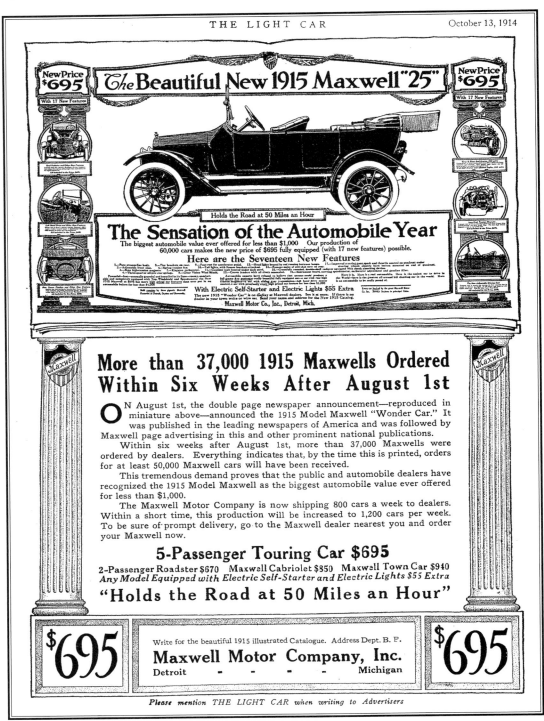

1915 was a good year – for Maxwell

Maxwells were manufactured in six different eastern cities. It was the forerunner of Chrysler in 1925. Maxwell achieved fame in 1909 when Alice Huyler Ramsey and three women friends travelled in a new Maxwell for their pre-Lincoln Highway trip from New York to San Francisco.

My father was a self-educated, pragmatic man from the farm, in a day when even high school was not the experience of most. Even so he possessed an innate artistic sense, evident in his photography. I treasure my father and his diary very much, but I also know it was not his way to describe with words the colors of the desert, how he felt at 20 below zero, or the people he met. He simply said, "cloudy" or "cold" or "fair" or "rain". When there was engine trouble he simply gave the facts about having to dismantle the engine to clean out the carbon. There is no mention of barked knuckles, dropped nuts and bolts, or lost washers. In fact he never said a word about what he ate on that long cross-country journey.

His economy with words was evident following his visit to the Exposition in San Francisco. He spent only one day there, including an evening in Chinatown. His comment: "Some sight!" The only other time he came close to exposing his feelings was in Elm Creek, Nebraska, on September 26, when he encountered the ubiquitous Nebraska mud. "Had H of time getting through," he wrote.

I remember once when I was a child and our family was driving between Downingtown and West Chester, Pennsylvania, that my mother commented on the beauty of a large oak standing alone in a pasture—its natural round symmetry and wonderful shade for the cattle grazing beneath it caught her attention. "Sure would make a lot of boards," was my father's take on it. Someone has hypothesized that every word spoken is stored in the ether waves forever. Perhaps sometime I can "click" onto the Lincoln Highway in 1915 and hear my father's Little Giant truck chugging across the dusty expanse of the Utah and Nevada deserts, or hear his verbal disgust at running out of gas at 10:30 p.m., alone and fourteen miles from any civilization (it could have been 100). Or hear his teeth chatter as he coped with below-zero temperatures in an open truck.

It could be argued that an ocean-to-ocean highway was first conceived in 1902, when the American Automobile Association suggested that such a road would be a good idea. In 1912 highway auto pioneers Carl G. Fisher, Henry B. Joy and F. A. Seiberling became involved in this developing pregnancy, giving final birth on July 1, 1913 when Joy became the first president of the Lincoln Highway Association.

Fisher's dream was for America's "Coast To Coast Rock Highway," to be paved in time for the 1915 Panama-Pacific International Exposition in San Francisco, at an estimated cost of $10 million. Many newspapers supported the idea: The Chesterton Tribune wrote on October 17, 1912:

The plan under present consideration is a big one, but with interest in it at a fever heat, and men behind it who know how to secure big results in an orderly and systematic way, in a few months, and the road itself an actuality by 1915, the time set by those whose ideas have gone into making a success of the venture. It is a project which is free from politics; the money for it is being subscribed by those who can afford it, and the completion of the road will result in benefit for the many. There is no class distinction here, and all encouragement should be given to the enterprise.

Seiberling, The president of the Goodyear Tire and Rubber Company said:

In my opinion, this move will succeed—it deserves to succeed, and the men behind it are calculated to see that anything they back gets its desserts. In other words, they are business men—and it takes business men, not politicians, to do all of the really big things of today.

Nissley Family Collection

1918 Saxon

My mother (on right) stands beside her 1918 Saxon. She would make us children laugh 'til our sides ached, telling about the time she hit a man on the Lincoln Highway going up Malvern Hill, steep even by today's standards. He was sprawled over the front fender all the way to the top because she didn't want to stop before reaching the summit. He crawled off and was no worse for wear and tear. Today there would have been two ambulances and six lawyers at the scene.

July 9, 1914 The Automobile Magazine
Road Smoothers

The Lincoln Highway enjoyed an upbeat reputation early on—at least for commercial advertising!

At a 1912 banquet in Indianapolis, Indiana, to promote the Lincoln Highway project, $300,000 was pledged by Seiberling after a thirty-minute presentation by Fisher. The primary appeal was made to the automobile industry, with another appeal to the general public. Certificates were made available from $5 to $1,000, with President Woodrow Wilson purchasing Certificate No. 1. The idea seemed to be on a roll. Unfortunately, by 1914 it became obvious that the highway would not be ready in time. Less than $4 million had been raised so the paving plan was quietly dropped. Perhaps if Henry Ford, instead of laughing, had supported the Lincoln Highway project it would have a different page in history.

Even so, the unpaved Lincoln Highway did begin at Times Square in New York, jogged down through New Jersey to Philadelphia, and went west through thirteen states to Lincoln Park in San Francisco. It traversed some 3,389 miles, depending on which Lincoln Highway jogs were taken, how often the wheels spun in mud or sand, or even in what year the trip was made.

The Lincoln Highway Association was a strong and viable organization for many years, even periodically publishing official road guides. However, the exploding popularity of the auto, given a huge boost with the advent of Henry Ford's revolutionary manufacturing process, resulted in the proliferation of named roads. Joy and Fisher were really onto something, with results far beyond their imagination. By 1925 there were at least 250 named highways, each competing for attention with their own signs and promotion. The resulting confusion inspired the federal government to launch a radical idea, a standard numerical system which still exists today: odd numbers, north-south; even numbers, east-west. The Lincoln Highway evolved into US 30 which more or less followed the Lincoln west from Philadelphia to Granger, Wyoming, where US 30 took off for Oregon.

By the end of 1927 named highways had disappeared and on December 31, 1927 the Lincoln Highway Association formally disbanded. Gail Hoag, its last field secretary arranged to have Boy Scouts install concrete marker posts from coast to coast on September 1, 1928 as a last hurrah. On February 11, 1939, when the second World's Fair was held in San Francisco, Hoag wrote to the highway supporters;

You know that your efforts helped to connect these two terminals (San Francisco and New York) by a completely paved road, and that our Lincoln Highway seedling miles are swallowed up in a network of pavement. The Lincoln Highway Association literally laid the foundation for a highway system which is the envy of the world.

By the 1940s most named highways were forgotten, and even public awareness of the well organized and publicity driven Lincoln Highway faced extinction. A generation was growing up with numbered highways and paved roads, and many had never even heard of the Lincoln Highway, but not everyone had forgotten. On October 31, 1992 in Ogden, Iowa, the Lincoln Highway Association was revived by Gregory Franzwa of Tucson, Arizona. The renewed association was now dedicated to preserving the highway by signing up people at the grass roots level, thus creating a conscience for its rightful place, not only in history, but in the travel plans and pleasure of the vacation-bound public.

YOU CAN GET A PACKARD IN SEPTEMBER

The season's production of Packard "2-38" six-cylinder cars was sold out in April and fell 500 short of market requirements. We thank our patrons for this endorsement.

For the past month we have been accepting orders for the next model—deliveries to be made in the early fall.

The Packard "3-38" contains all the essential features of the previous model, and in addition those refinements which would naturally accrue at this advanced stage of Packard development.

Twenty styles of open and closed bodies, ranging from two to seven passengers, give a wide choice, fitted to every use and satisfying the individual taste.

There will be no advance in price. Catalog on request.

PACKARD MOTOR CAR COMPANY
DETROIT · MICHIGAN

Lincoln Highway Contributor

Licensed Under Kardo Patents

Packard "3-38" Standard Touring Car, Seven Passengers, Price $3,850

Please mention The Automobile when writing to Advertisers

July 19, 1914, The Automobile Magazine

1915 Packard

The 1915 Packard Twin Six was an important car driven by Henry B. Joy and Austin Bement to San Francisco in 1915—the same year my father drove his 1915 Little Giant to San Francisco.

TWO REASONS WHY DISCRIMINATING PEOPLE EXPLORE THE LINCOLN HIGHWAY IN EASTERN PENNSYLVANIA

Pennsylvania Dutch Farms

Along Route 23 just over the ridge from the Lincoln Highway

Amish Cornfield

Along Rockford Road just south of the Lincoln Highway

PENNSYLVANIA
And The Lincoln Highway

Miles: 362 , Towns: 80 (1924 Guide)

PENNSYLVANIA OCCUPIES A HIGHLY significant, and I fear under-appreciated, role in the annals of Lincoln Highway history. The concern for paved highways in Pennsylvania predates the official Lincoln Highway by 118 years. In 1795 the Philadelphia to Lancaster Turnpike was officially opened—it could be argued, the first paved highway in the nation. If there is ever a Lincoln Highway Hall of Fame, Pennsylvania should be prominently represented.

With the advent of the auto around 1900, there emerged a growing concern for better roads. In many places prior to 1900, horses, buggies, and wagons were adequate for local transportation, and trains for town-to-town travel. This was not the case in Pennsylvania. Concern for better roads goes back to colonial history, when Lancaster was the provisional capital of the American Colonies and Philadelphia the provincial capital. As early as January, 1730, magistrates, grand juries and other inhabitants of Lancaster County presented a petition to the Provisional Council, setting forth

> ...that not having the Conveniency of any Navigable water, for bringing the produce of their labours to Philadelphia, they are obliged at a great expense to transport them by Land Carriage, which Burthen becomes heavier thro' the Want of Suitable Roads for carriages to pass. That there are no public roads leading to Philadelphia yet laid out thro' their County, and those in Chester County, thro' which they now pass, are in many places incommodious.

By 1772 the Pennsylvania General Assembly had passed an act calling for the construction and maintenance of public roads and highways. Thomas Mifflin, Pennsylvania's first governor, formed a study committee that in February, 1772, reported on the need for an artificial road to be funded by the public. In March 1772 a bill was introduced directing how the road would be built and maintained. It contained twenty-two articles. Shares of stock would be sold for $300 each. The response and sales were a phenomenal success. Two men—Thomas Telford and an Irish engineer, John McAdam—devised a method to cover the road bed with coarse rock topped with finely crushed stone with a center crown, making the road passable year round. The new road would be called, the "Philadelphia-to-Lancaster Turnpike." It would be 62.5 miles long, beginning at the west bank of the Schuylkill River and ending in the center of Lancaster. It would be fifty feet wide with twenty-one of those feet covered with gravel. Estimated cost: $464,142.31

The highway was officially opened in 1795, but not completely finished until 1796. Investors received dividends collected from the nine tollgates. This new, innovative and strategic road became the first long-distance, hard-surfaced road in the country and in 1913

Lincoln Highway Remnant, 2002

Immediately beyond the former PRR underpass, the old Lincoln Highway made a sharp left turn to parallel the railroad. Now it is used only as an access to a few private homes near Malvern.

Haldeman Farm House, 1965

Here, in Malvern, PA, is where my wife lived on the Lincoln Highway as a child, and where George Washington stayed one time with his generals. It was also the location of the 1795 "21 M to P" marker. In 1977 this house became a practice burn for the local fire company, to reduce taxes.

became an important part of the Lincoln Highway. It is interesting to note that twenty-three years transpired between the original conception in 1772 and its birth in 1795. Carl Fisher and Henry Joy hoped to do the length of the entire country in less than two years. The Philadelphia-Lancaster Turnpike was a combination of politics and private enterprise, whereas in 1912 Henry Ford insisted on tax funding, with Fisher and Joy preaching the merits of private.

The positive effects of this 1795 road are incalculable. Not only was commerce enhanced by transportation, but ancillary businesses such as inns, restaurants, and housing prospered as well. The most significant benefit was societal. This road tied communities together and facilitated family and religious communication. I, too, have been a beneficiary of this long-ago event. In the consciousness of my wife and myself, the Lincoln Highway has always existed. My wife lived just a quarter block away in Paoli, Pennsylvania, when she was a young child. My family moved to close-by Valley Forge when I was two. Later we both worked in my wife's family-owned farm markets on the Lincoln Highway in Villanova, Wayne, and Malvern. When my mother was 16 she lived with her family in a haunted house—the famous (or infamous) Blue Ball Inn in Daylesford, Pennsylvania on the road that later became the Lincoln Highway. Even though our fortunes as adults have taken us far afield, we always considered the Eastern Pennsylvania Lincoln Highway environs "home." Come visit us and we will invite you to sit on a stone bench which graces the entrance to our front door. In its former life, it was a 1795 "Lincoln Highway" mileage marker, "21 M TO P" (21 miles to Philadelphia) on the farm where my wife grew up. Conversations are now in process for its "reincarnation" to its original sentry post. This will be a historical closure event.

Haunted House

This house is on Conestoga Road, formerly the Lincoln Highway, at Daylesford, Pennsylvania, about eighteen miles west of Philadelphia. The story begins with Prissy Robinson, who owned and operated the inn from 1799 to 1831. She would kill traveling salesmen for their money and wares, then bury them in the dirt floor of the basement or back in the orchard. Many stories were told by my mother, who lived there with her family when she was sixteen – blood stains on the floor, doors which would not stay shut and loud, unexplained crashing sounds.

21 Miles To Philadelphia

However, some things, it seems, never reach closure. A long driveway and frolicking lawn away from this marker stood the stately old stone farmhouse of my wife's parents, constructed during George Washington's era; in fact, George himself once stayed there with his generals. It was a three story stone house with fireplaces upstairs and down. Unfortunately, in 1977 it became a practice burn for the local fire department—a victim of high taxes.

So much for historical preservation.

This section of the highway today is deluged with not only local commuters but with millions of tourists drawn annually to the beautiful farmland and cultural uniqueness of the area, one of the leading tourist destinations in the United States. In addition there are long lines of eighteen-wheelers, in spite of the parallel Pennsylvania Turnpike, just a few miles to the north.

The Lincoln Highway between Philadelphia and Lancaster holds some of the most significant transportation and commerce history in the United States. If it could speak, what stories it could tell! From meandering Indian trails to mud roads to gravel to concrete, and today, too often, to an unsightly sixty-five-mile string-town.

The time is past due for a new committee to research the viability of renovating this national treasure. A highly significant watermark in history has already passed us by—1995, the 200th Anniversary! May I be audacious enough to propose the creation of a corridor park along the Lincoln Highway from Philadelphia to Lancaster? As the Philadelphia-Lancaster Turnpike in 1795, was the first long-distance paved road in the country, the *Lincoln Highway Corridor Parkway* could be the first historical park of its kind in the country—a sixty-five-mile long "Williamsburg," if you please. Too grandiose? Shades of Fisher and Joy and their plan to pave the 3,000 miles of Lincoln Highway from the Atlantic to the Pacific for $10 million in just two years! But that didn't work out either, did it? Perhaps a strategically placed "Seedling Mile" might work.

Philadelphia-Lancaster "Turnpike"
at the turn of the twenty-first century

Looking eastward on the Lincoln Highway near Malvern, Pennsylvania. The view of congested traffic and commercial, hodge-podge signage speaks for itself.

Gap Hill, 2002

Looking westward on the Lincoln Highway to the "Garden Spot" of Lancaster County, from Gap Hill.

The New SAXON $395

What the SAXON Coast-to-Coast Trip Over the Lincoln Highway Means to You

A Saxon car traveled 3389 miles overland from New York to San Francisco in 30 days across the Lincoln Highway.

It was the first automobile to make a continuous trip from New York to San Francisco over the Highway and the first car of its size and price to make the journey from coast to coast. It averaged 30 miles to the gallon on the long trip.

The same car, before starting on its transcontinental trip, ran 135 miles a day for 30 consecutive days—4050 miles—averaging 30 miles to the gallon of gasoline and 150 miles per quart of oil, covering the entire distance on the original set of tires.

In 60 days this car covered almost 8000 miles—as far as the average owner drives in two years.

What this Proves to *You*

It proves that the Saxon is a two-passenger automobile equal to any test which any user would ever give it. A car that can fill every need for business, for health, for pleasure—a car everyone can afford to buy and to keep.

6000 satisfied owners all over this country are using Saxons and proving the economy and utility of these sturdy cars every day.

The Saxon car today, is the best two-passenger automobile in the world at anywhere near its price.

The new Saxon with running-boards and 38 other improvements is selling fast throughout the country.

You will find it to your advantage to write us regarding open territory. Why not do it today? Address Dept. Z.

Saxon Motor Company, Detroit

October 13, 1914, The Light Car Magazine

A Lincoln Highway Pioneer —Saxon

Fifty some years before Rambler coined the phrase, "Compact Car", the Saxon epitomized the concept of the "Light Car". Even then there existed a niche market for small cars. Saxon, however, got a leg-up on other contemporary light cars by being the first auto (1914) to travel the newly designated Lincoln Highway-3,389 miles in just thirty days, averaging thirty miles per gallon. That's credible even today.

Birth, Demise and Rebirth of Lincoln Highway Posts

Since 1745, *the historic General Warren Inne has been center stage on the pre-Lincoln Highway near Malvern, Pennsylvania. The name changed over the years depending on owner or politics. Its been known as: Admiral Vernon (first name), Admiral Warren (1758), and General Warren Inne (1825). It was a Loyalist/Tory stronghold during the Revolutionary War and is now a perfect blend of oldworld charm and modern accommodations. It sits just south of the Pennsylvania Railroad on the old Lincoln Highway.*

ON SEPTEMBER 1, 1928, THE BOY SCOUTS of America set 2,346 (The credible count by Russell Rein, a long time Lincoln Highway enthusiast and first fiddle when it comes to Lincoln affairs) concrete posts along the Lincoln Highway from coast to coast. It stretches the imagination to envision Boy Scouts planting 7-foot-long concrete posts, almost 3,000 of them, and each one much too heavy for one person and a bit much even for three. But then Boy Scouts were and are a resourceful bunch.

During my early scrutiny of the Lincoln Highway, my informants indicated that only about twenty-four still remain. Obviously they were ill informed. There are that many in Pennsylvania alone and over eighty in Iowa, but even so its only a fraction of what once was. There is a story in Iowa of a highway crew with a road renovation assignment who dug a hole, buried 100 posts and filled the hole with concrete. Some were thrown into rivers, some just thrown away, and some as foundation for bridge abutments in Columbus, Nebr. Some became road kill by the very cars they were dedicated to serve. There was a time in history when a Lincoln Highway post was considered worthless, in the way, a has-been. Happily, there is a resurgence of interest and these silent sentinels are being resurrected.

A man bought one at a flea market for $40. Later his son received a $4,000 tax credit by donating it to the Fish Springs Wild Life Refuge on the Lincoln Highway in Utah. A mother somewhere along the route got tired of tripping over this thing in her basement and instructed her teenage son to take a sledge to it and get it out of the way. Sometimes procrastination pays off as in this case. The son came down with a serious case of "I'll-do-it-later-itis." One day conscience motivated him enough to at least turn the thing over, revealing the medallion, but even then he delayed action until one day he viewed the tail-end of a Public TV program featuring the Lincoln Highway—a post was shown. "Hey, we have one of those," he exclaimed, and subsequently the post found its way to a good home. A Lincoln Highway neighbor once saw a post being manhandled by a road repair crew and literally rescued it from the grave, even though it was badly wounded. It awaits constructive surgery in the office of the Pennsylvania Lincoln HighwayHeritage Corridor, then in Greensburg, under the tender care of Olga Herbert. Posts are being resurrected from barns, ditches, sheds and dumps across the country—some many miles from where once they so proudly directed tourists to the right, or left . The LH posts helped bring an end to dependence on railroad routes to get from A to B.

The following is an incomplete list of identified concrete posts in Pennsylvania with directions to their locations. Locating them and taking their pictures has been a lot of fun, even though the quest was often under adverse circumstances like rain, intrusive traffic, or darkness. Some were surrounded by uncomplimentary environment, many needed a flash and some a telephoto lens to take a shot from across the street between whizzing cars. If you want something exciting to do tomorrow, go look for a post! When you do, please remember that many are on private property. so please be respectful and do not disturb.

Malvern, Pennsylvania

Malvern, Pennsylvania on the old Lincoln Highway one mile south of route 30 across from the Villa Maria Academy.

Courtesy of Kenneth Leasa

Hallam, Pennsylvania

Located in the Kreutz Creek Valley Preservation Society, 5345 Lincoln Highway, Hallam, Pennsylvania.

Leaman Place, Pennsylvania

Southside, west of the railroad, down a steep bank behind the guardrail. This post was removed during the mid 90s bridge renovation and replaced by the Pennsylvania DOT at the present location.

York, Pennsylvania

1501 East Market, northeast corner of Ogontz and East Market streets.

Columbia, Pennsylvania

North side of 311 Chestnut Street on Highway 462. Note the very old brick sidewalk.

York, Pennsylvania

1341 West Market Street, north side of street next to West York Fire Station.

Hellam, Pennsylvania

South side of the road across from Hellam Township Building, east of Hallam. Note: "Hellam" is the township and "Hallam" is the Borough. Early instruction for finding this post said, "across from the Wise Owl". We looked in vain for the "Wise Owl" restaurant, motel or some such. When we explained our dilemma to the local police chief, he pointed to a Wise Owl potato chip sign along the road next door.

Thomasville, Pennsylvania

Southwest corner of North Grant and Lincoln Highway, across from Martin's Chips offices.

New Oxford, Pennsylvania

In the town square on the west side.

Chambersburg, Pennsylvania

Northwest corner of Lincoln Way and 3rd Streets.

New Oxford, Pennsylvania

South side, 1.7 miles west of the town square.

Chambersburg, Pennsylvania

Northwest corner of Lincoln Way and Franklin streets - one block east of Federal Street

Gettysburg, Pennsylvania

On second floor of the Adams County Historical Society museum in Schmucker Hall, on the seminary campus.

Fort Loudon, Pennsylvania

West of Fort Loudon, 2 miles west of bridge on the curve to the right. North side of the road

Stoufferstown, Pennsylvania

2.3 miles east of Chambersburg on south side of street.

McConnellsburg, Pennsylvania

In front of historic Fulton House, 112 Lincoln Way, east on north side of the street.

Harrisonville, Pennsylvania

North side of the road west of a crossroad in a private yard. The post was rescued following an encounter with a snow plow. In the Spring it is surrounded by more than 100 tulips.

Buckstown, Pennsylvania

On south side of the road across from Ridge Road, buried to its neck. Note that the medallion is missing.

Everett, Pennsylvania

Lincoln Highway Heritage Corridor sign two miles east of Everett. Note the Allegheny Mountains.

Stoystown, Pennsylvania

On south side of the road 1/2 block east of Kantor Street at the curve.

Everett, Pennsylvania

Northwest corner of Hillside Street and Route 30, just before the curve to the right.

Ligonier, Pennsylvania

Northeast corner across the street from the town square.

Bedford, Pennsylvania

East of intersection of Routes 30 and 31 on north side of the road, just east of the Jean Bonnet Tavern.

Youngstown, Pennsylvania

Just .8 mile west of traffic light, on southwest corner of Club Manor road and Route 30

Greensburg, Pennsylvania

521 East Pittsburgh Street on the northeast corner of East Pittsburgh and Fremont streets at Barnhart Funeral Home. This is a remanufactured post.

Greensburg, Pennsylvania

Olga Herbert holds a LH post rescued from a highway renovation project

Does The Old Geezer Still Drive?

"THE THING THAT KEEPS YOUR FATHER FROM HAVING accidents is that he's so quick," was my mother's explanation for my father's impeccable driving record. To my recollection his first accident was at the age of sixty-four when he totaled my '33 Chevy by driving in front of a milk truck. It was no contest.

At age 85 he volunteered to hang up his car keys following several close calls in his daily travels into Philadelphia from Phoenixville, Pennsylvania, where he was living. My family lived in Goshen, Indiana, at the time (just off the Lincoln Highway), when we became the custodians of my father's 1958 Rambler American station wagon. As the preacher said, "Mortality is 100%." So also are the limits of our driving. Death by motor vehicle is a leading cause of death in the United States—one person every 13 minutes. The following are a few suggestions for diagnosing the level of safety for ourselves and others:

Do you have trouble seeing lines and other pavement markings, curbs, medians, other vehicles and pedestrians, especially at dawn, dusk, or at night?

Do you experience more discomfort at night from the glare of oncoming headlights?

Do you have trouble looking over your shoulder to change lanes or looking left and right to check traffic at intersections?

Do you have trouble moving your foot from the gas to the brake pedal, or turning the steering wheel?

Do you walk less than one block a day?

Do you feel pain in your knees, legs or ankles when going up or down a flight of stairs?

Do you feel overwhelmed by all of the signs, signals, road markings, pedestrians, and vehicles that you must pay attention to at intersections?

Do you take medications that make you sleepy?

Do you lack confidence that you can handle the demands of high speeds or heavy traffic?

Has a friend or family member expressed concern about your driving?

Have you had several moving violations, near misses or actual crashes in the past three years?

Susquehanna River Bridges

Between Columbia and Wrightsville

Ferry
    1733—John Wright, Jr. secured license to operate a ferry.

First Bridge
    1814—5,690 feet long, wood covered
    1832—February, bridge destroyed by ice/flood

Second Bridge
    1834—5,620 feet long, wood covered
    1863—June 28, bridge burned down by Union forces,
    a prelude to Gettysburg

Third Bridge
    1868—5,390 feet long, including railroad, wood covered
    1869—January 4, bridge opened – two metal spans in the middle
    1896—September 30, bridge destroyed by a hurricane.

Fourth Bridge
    1897—5,300 feet long on 27 piers, steel covered bridge
    1958—Discontinued use
    1963-64—Dismantled

Fifth Bridge
    1930—7,500 feet long, concrete. 25 cents toll

Sixth Bridge
    1973—5,634 feet long, Route 30 bypass

COLUMBIA / WRIGHTSVILLE
Pennsylvania

COLUMBIA AND WRIGHTSVILLE SHARE A COMMON HERITAGE separated only by the great Susquehanna River. This river holds a very personal interest for me because my father grew up on a farm just a few miles east of the river near Washington Boro, and I was born a few miles north of Columbia/Wrightsville in Lemoyne, Pennsylvania, on the bluff overlooking the Susquehanna River.

COLUMBIA

IN MY FATHER'S OWN WORDS:

Saturday, June 19, 1915

In camp all day. Went to Columbia in p.m.

Wednesday, June 23, 1915

Booked Dreamland Theater 23 & 24 in Columbia 50% - 50% [division of theater receipts]. Received $18.00

Thursday, June 24, 1915

Booked Dreamland Theater. 50% - 50%

WHY ARE TOWNS BORN? For some towns, it's someone's dream come true. For some, it's circumstances, or a river, or a railroad, or a valley, or an event, and for some it's a person. Such was Shawanatown. The person was a Quaker, John Wright. The year was 1724 when he came to preach to the Shawnee Indians settled along close by Shawnee Creek.

John Wright, typical of most people who find a good thing, shared it with others. In 1726 Robert Barber, a friend of Wright's, moved in and built a log home and sawmill. Wright himself and another Quaker friend, Samuel Bluston, arrived later.

Columbus Historic Preservation Society

Burning Bridge, 1863

An artist's conception of the second bridge's demise from an authentic Civil War engraving. The original sketch was by A. Berhaus. On June 28, 1863, it was set afire to halt the Confederate advance on Lancaster. The bridge was 5,620 feet long, 40 feet wide on 27 pillars, and cost $128,726.

Hurricane Bridge, 1869

The third bridge *(covered) was 5,390 feet long with 29 spans built upon the piers of the previous bridge at a cost of $400,000. To protect it from a possible future fire, two iron spans were placed in the middle. What they did not count on was another vagary of nature – wind. On September 30, 1896 this third bridge was destroyed by a hurricane.*

Railroad Bridge, 1897

The fourth bridge *(covered) was 5,300 feet long on 27 piers and cost $455,000. Prior to the fire disaster in 1863, the bridges were owned by the Columbia Bank. After the smoke cleared and the timbers and war cooled down, the bridge was sold, and the Pennsylvania Railroad became next to invest in a Susquehanna River bridge. Thus it was that the 1897 bridge became the "Railroad Bridge." Since the three previous bridges had been destroyed by ice, fire, and wind, the railroad wisely decided on steel for the 1897 bridge. The original intent was for a two-deck affair with trains on the bottom deck and wheeled vehicles on the upper. In reality the upper deck was never finished so cars and trains shared the bottom level. This is the bridge my father traveled in 1915.*

Concrete Bridge, 1930—Looking west from Columbia, 1994

The fifth bridge, *(concrete) looking west from Columbia, replaced the former steel bridge to accommodate the demands of increasing auto traffic, with a cost of $2,484,000 (toll, 25 cents per car). Following the withdrawal of auto traffic on the steel railroad bridge, passenger and freight trains continued using the bridge until January 1954, when the bridge was closed due to declining railroad use. In 1963 the bridge was dismantled for scrap metal just 100 years after the burning event leaving just a few naked piers. Note a remnant pillar and the 1973 bypass bridge to the far right..*

1930 Concrete Bridge

The fifth Bridge; *The 1930 concrete bridge as viewed from the 1972 four-lane Route 30 bypass bridge. This 7,500 foot long, 28-arch bridge, is still in use connecting downtown Wrightsville with Columbia. Note the remnant piers of the bridge my father traveled in 1915, clothed now with moss, vines, small bushes and trees.*

Since Shawanatown was the lowest point along the Susquehanna River where it could be crossed during low water, John Wright took out a patent for a ferry in 1730 and changed the name to Wright's Ferry. It is said that the ferry became so popular that at times people would have to wait two days to cross. For the next eighty-four years people crossed the river via Wright's Ferry, until such travel abruptly ceased in 1814 with the construction of a covered bridge.

This growing pioneer community along the Susquehanna River became a family affair. Not only did John Wright provide the dream, the example and the ferry, but his daughter, Susannah, was no slouch either. The Wrights came from England, and Susannah assumed the role of mother for her younger siblings when their mother died. This in itself was not unusual for older children in pioneer families. What was unusual was her proficiency in the arts, needlework, landscape painting, medicine, and law. In fact, she was honored by the queen of England for raising silkworms and thus making Columbia a center for silk manufacture.

Following Susannah's death at age 84 in 1785, her estate went to a nephew, Samuel, who had the land surveyed, and laid out in 160 building lots. He sold chances by lottery and called the town Columbia, after Christopher Columbus. Columbia later came within just two votes of being chosen the nation's capital.

In the early sessions of Congress in 1778 and 1779 considerable time was spent on the future permanent seat of the federal government. Several sites, including Columbia, were under consideration. George Washington, among others, favored Columbia. Among the strongest proponents was Pennsylvania Senator William Maclay who passionately argued for Columbia at every opportunity. The final vote was taken during a dinner at the home of Thomas Jefferson and the spot on the Potomac River won out over Columbia, on the Susquehanna River, by only one vote. Maclay said he knew nothing of the Jefferson dinner so was not present to push the issue. The present site of Washington, D.C., was a concession to southern states. Can you imagine the interesting changes in Lincoln Highway history if Columbia had become our nation's capital?

Miriam stands beside the lone remaining post in Columbia. It is at 311 Chestnut Street, north side, on route #462.

A January View of the Susquehanna River, 2002

A view of the Susquehanna River looking south from the Columbia side of the concrete bridge.

Keeley Stove Truck, 1915

The Keeley Stove Company's truck in Columbia, Pennsylvania, 1915. My father may have had to move over to let this truck pass.

WRIGHTSVILLE

IN HIS OWN WORDS:

Wrightsville, PA. June 12, 1915

Fair, made delivery in Marietta and moved to Wrightsville, Pa. Received $4.22, ps $2.00 [from sale of photos in Marietta.]

June 13, 1915

Rain. Booked Wrightsville for Thurs. and Fri. 24th & 25th. Went down home, [less than ten miles on farm near Washington Boro] with John [his older brother]. Rained all day.

June 14, 1915

Fair. Made 14 shots [his words] in Wrightsville. Developed [photographic] plates.

June 15, 1915

Rain. Made 25 shots in Hallam and 20 shots in Wrightsville.

June 16, 1915

Fair. Made a few shots in town and went to picnic at Chick Park in P.m. Got stuck in bad road.

June 17, 1915

Fair. Developed plates for picnic etc. and advertised for show.

June 21, 1915

Delivered in Wrightsville and left for Red Lion, Pa. Good roads

OVER THE YEARS THE SUSQUEHANNA RIVER became the focal point for Columbia and Wrightsville. It provided a natural, maybe even romantic, setting for Wright's Ferry, the famous covered bridge, and in 1826 a canal built with a $300,000 appropriation from the state. This thrived until one day a railroad locomotive arrived pulling three passenger cars. The canals gradually gave way to railroads which dominated the transportation scene in the United States

Columbia Historic Preservation Society

Roundhouse, c. 1900

The Pennsylvania Railroad Roundhouse in Columbia, when train traffic through Columbia/Wrightsville was at its height. Only remnants remain. This is a sight probably common to my father in 1915 along with other travelers of the Lincoln Highway.

Lyric Theater, Wrightsville, Pennsylvania, 1915

The theater my father saw when he booked it for June 24 and 25, 1915 to show off the children from Wrightsville and Hallam on the screen. Actually the date was changed to June 17 and 18.

Lyric Theater, Wrightsville, Pennsylvania, 2002

The Lyric Theater eighty-five years after my father booked it in 1915.

until the days of Dwight D. Eisenhower and the interstate highway system. Eisenhower, by the way, got his first hands-on experience with bad roads in 1919 as a young lieutenant with the military truck convoy over the Lincoln Highway to San Francisco. Later when United States president, he initiated the interstate highway system. It seems only fair that the Lincoln should receive same credit by predating Eisenhower by six years in 1913.

Wrightsville and Columbia occupy strategic positions on the Susquehanna River where, over the past 250 years they played an important role in both the Revolutionary War and the Civil War. For our purposes, they served as a proving ground for the westward wagon flow of earlier western settlers and later the flood of autos over the Lincoln Highway. Speaking of railroads, the Underground Railroad had a secret station in Columbia for the passage of an estimated 40,000 runaway slaves.

The history of Wrightsville began as a part of York County history, with agreements between William Penn and the Indians. In 1682 Penn and his heirs signed a peace treaty with the Onondaga, Seneca, Oneida, and Tuscarora nations, deeding "All the river Susquehanna and all land lying on the west side of the river to the setting of the sun," to Penn and his heirs.

The most important historical artifacts for these two communities were the bridges connecting them across the river. Of these, the most dramatic was the second, built in 1834, which burned in 1863. The events leading up to this tragedy changed the course of history for both towns and for us.

Late in June 1863, Confederate troops spread across York County as far as the Susquehanna River in Wrightsville. Union forces fled into Lancaster County, and further advances of the Confederates were checked by an order from the Union militia. Four Columbians: John Q. Denny, John Lockard, Jacob Rich, and John Miller set fire to the bridge to halt Confederate advances. Their intent was to burn only enough to make it impassable, but the fire got out of hand and destroyed the entire bridge. This single event resulted in the famous, or infamous, battle of Gettysburg. And, as they say, the rest is history. Fifty years later, in 1913, the site of this bridge became the Lincoln Highway.

Modern Columbia and Wrightsville are nostalgic by-ways no longer throbbing with canal traffic or puffing, steaming, smelly, tszzzzzing locomotives, or manufacturing. They do offer a wonderful view of history through their museums and parks and the small-town charm in western Lancaster County and eastern York County. They are an excellent one-tank trip for anyone in eastern Lincoln Highway country.

An Endorsement, 1915

The above unsolicited letter of endorsement from H. Ogden Birnstock, proprietor of the Lyric Theater in Wrightsville, was found among my father's many boxes of photographs and memorabilia.

GETTYSBURG, *Pennsylvania*

IN HIS OWN WORDS:

Gettysburg, Pa., July 1, 1915

Fair. Landed in Gettysburg. Camped in furniture factory lot. Roads good. Booked 7th & 8th at Walters Theatre, Gettysburg, Pa.

Saturday, July 3

Fair. Shot 25 in a.m. for show Wed. & Thurs at Walter's Theatre

Wednesday, July 7

Fair. Passed proofs and shot factories for Friday night show. Came back and showed in Gettysburg. $69.70

Thursday, July 8

Cloudy, rain. Delivery and proofs of Gettysburg Show receipts tonight $46.10

Friday, July 9

Fair. Run off kid orders for Gettysburg. Delivery $26.20

Saturday, July 10

Fair. Passed proofs and made delivery in Gettysburg.

THE HISTORY OF GETTYSBURG and its environs is spread over two centuries going back to Indian raids and courageous pioneers seeking a new life in a new land on a new frontier. But the area is best known for just three days of that time, July 1, 2 and 3, of 1863—the Battle of Gettysburg where 51,000 men were killed, wounded, or listed as missing. Much grass has grown since and flowers have bloomed over those hills, as though to mask the horror of those three days. It has been more than a century now, but not even time can erase the suffering, pain, and lost innocence of a nation at war with itself. Nearly two million people visit the site annually to view the 1300-plus monuments and 360 cannon spread over forty square miles.

In 1736 the land was purchased from the Iroquois Indians by the family of William Penn. In just a few years 150 families of Scotch-Irish descent settled in the area which became known as Marsh Creek Settlement.

By 1761 Samuel Gettys, one of the Marsh Creek settlers, established a store and tavern, but nothing could be called a "town" until 1786, when Gettys' son, James, laid out 210 lots centered on a square that was laid out at the intersection of two roads. This town became known as Gettysburg in the county of York. With the approach of the nineteenth century, the growing population of the town chose to separate from York County by creating a new county named after President, John Adams with Gettysburg serving as the county seat.

By 1860 Gettysburg had grown to a population of 2,400 persons with ten roads leading into it, two of which fifty-three years later became the Lincoln Highway. At the time of the Battle of Gettysburg there were about 450 buildings housing carriage manufacturers, shoemakers, tanneries, banks, and taverns, along with an assortment

A Little Giant At Devil's Den, 1915

My fathers' crew and truck stand by their cameras at Devil's Den, Gettysburg. My father took the picture.

of the usual retailers.

In 1915, on his way to the Panama-Pacific International Exposition in San Francisco, my father spent ten days in Gettysburg. This adventurous journey over the coast-to-coast highway in those days was the privilege of those affluent people who had enough resources to purchase the car, spare tires, and fuel and to make the necessary repairs on the trip, to say nothing about time off from work. For my father, all it took was a vision and sufficient pioneering pluck to pull it off. He was thirty-three, single and a traveling professional photographer of children. During those ten days in Gettysburg he camped on the Reasor Furniture Company factory lot, located in the northeast part of the town, north of the Western Maryland Railroad. He booked the Walters' Theatre for July 7 and 8, and photographed as many children, factories, and businesses as he could talk into it, for public showing on the screen.

History Repeats Itself

Some eighty-five years after my father parked his Little Giant at Devil's Den on his way to San Francisco, my wife and I parked our Subaru within ten feet of my father's track. We, too, were on our way to San Francisco following a small leather-bound, stained diary.

TRANSCRIPT OF THE GETTYSBURG ADDRESS ON DISPLAY AT THE NATIONAL ARCHIVES.

FOUR SCORE AND SEVEN YEARS AGO OUR FATHERS BROUGHT FORTH, UPON THIS CONTINENT, a new nation, conceived in liberty, and dedicated to the proposition that all men are created equal.

Now we are engaged in a great civil war, testing whether that nation, or any nation so conceived, and so dedicated, can long endure. We are met on a great battlefield of that war. We come to dedicate a portion of it, as a final resting place for those who died here, that the nation might live. This we may, in all propriety do. But, in a larger sense, we cannot dedicate, we cannot consecrate, we cannot hallow, this ground. The brave men, living and dead who struggled here have hallowed it far above our poor power to add or detract, The world will little note nor long remember what we say here, but it can never forget what they did here.

It is rather for us the living, we here be dedicated to the great task remaining before us – that from these honored dead we take increased devotion to that cause for which they here gave the last full measure of devotion – that we here highly resolve that these dead shall not have died in vain, that this nation shall have a new birth of freedom, and that government of the people by the people for the people, shall not perish from the earth.

Library of Congress exhibition

The American President

Lincoln was particularly proud of his height – six feet four inches in his stocking feet – and he made political hay of it whenever he could, either through humor, or through challenges to other tall men to measure against him, or by looming over those he debated or disagreed with.

The earliest known likeness of Lincoln was taken in 1846.

By the time Lincoln was forty-five, his face had lost its roundness.

He could look handsome and bland, as this ambrotype image shows.

He could also look horsy and ruthless, as he did here in 1860.

Beardless, his face could spark with intensity.

With whiskers, in 1861, the same face seems muted.

In need of a haircut, Lincoln has just arrived in Washington.

By late 1863, two and a half years of war have aged him.

Lincoln had a drooping eye that tended to make him look tired.

In this 1864 profile Lincoln still looks vigorous.

It may be that Lincoln had his hair cut short to prepare for a life mask.

In his last year of life, at fifty-six, Lincoln looked like an old man.

Lincoln was caught by at least thirty-one camera operators on at least sixty-one separate occasions, making him the first heavily photographed president in American history. All the pictures before he was elected show him clean-shaven; in all the pictures after his election he wears a beard. Most of the time he looks stern or melancholy, not letting his face sparkle as it often did. Friends said no photograph ever really captured him, that the complex movement of his facial muscles was so much a part of his looks that to freeze them only diminished him.

Walter's Theatre 1926

My father booked this theater on July 7 and 8, 1915, to show off the children he "shot" (his words) and local businesses, on the screen. Eleven years later the theater metamorphosed into a garage for Hudson and Essex cars, or whatever came along. Note the cost of a new Hudson or Essex.

WALTER'S THEATRE
TO-NIGHT
YOUR LAST CHANCE

To see yourself and friends on the screen. Everyone who saw the pictures last night had only words of praise for them.

A GREAT PICTURE PROGRAM

MR. BUTTLES..ESSANAY COMEDY
   IN THREE ACTS FEATURING RICHARD C. TRAVERS and EDNA MAYO
ROSELYN...VITAGRAPH DRAMA
   IN TWO ACTS FEATURING NAOMI CHILDERS.
ADMISSION 10 C   CHILDREN 5 C

Walter's Theatre Ad

Newspaper ad in the Gettysburg Times July 8, 1915, promising entertainment plus showing local children and businesses.

Walter's Theatre 2002

At 26-28 East York Street in Gettysburg, the Walters Theatre wears its turn-of-the- 21st century-face.

Lonesome Post

On second floor of the Adams County Historical Society museum in Schmucker Hall on the seminary campus resides the only remaining post, that I could find in Gettysburg.

CHAMBERSBURG, *Pennsylvania*

IN HIS OWN WORDS:

Wednesday, July 14, 1915

Fair. Left for Chambersburg, Pa at 5 A.M. Good roads. Booked 2 days at Orpheum Theatre Camped at Wolf Park.

Tuesday, July 20

Fair. Run slides for show at Orpheum.

Wednesday, July 21

Fair. Had $59.00 for house in Chambersburg, Pa.

MODERN CHAMBERSBURG is an enthusiastic thriving town of 17,000 persons within 100 miles of the much larger and internationally prestigious communities of Baltimore and Washington, D.C. It is host to the intersection of two major U.S.. highways: U.S. Route 11 and the venerable Lincoln Highway, U.S. Route 30. The Lincoln Highway in Chambersburg also hosts the *Old Franklin County jail,* built in 1818. It is the oldest jail in Pennsylvania and the longest continuously operated jail, being officially used from 1818 to 1971. Within the exercise yard to the east of the cellblock stood the gallows for public viewing of executions. For spectators a hanging was not only a holiday, it was a "scaffold service." Right triumphed over wrong. It currently is home to the Kittochtinny Historical Society.

Other historical sites my father may have seen were: *Fort Benjamin Chambers*, a personalized fort built by Benjamin Chambers in the winter of 1755–56 for protection against possible Indian attacks, and *White Squaw Marker*, a memorial to the abduction of Mary Jamison in April 1758 by a party of Indians and Frenchmen. Mary's family was scalped but she survived and married within the tribe. The marker gives an intriguing narrative of the event. *Fort Loudon Site* was built by the British in 1756 as a superior defense during the French and Indian War and was also the site of first American Revolution casualties in 1765—ten years before Lexington or Concord. Also *Wannamaker Row,* a row of townhouses built

Chambersburg Jail, 2002

In 1818 the Franklin County jail in Chambersburg became the strongest jail in the state, with three-foot-thick limestone walls twenty feet high (see below) with only two large double door openings. One cell in the basement had its only entrance through another cell framed with rough hewn timbers and was reportedly used for hiding escaped slaves. That was a perfect place for feeding and security until they could safely continue their journey northward to Canada.

by Nelson Wannamaker, father of the renowned John Wannamaker, from his own local brickyard, and *152 Lincoln Way East*, where Dr. Benjamin Seseney produced the first commercial smallpox vaccine in the world.

During the Civil War, Chambersburg became one of the many unfortunate casualties of that dark chapter in American history. In 1863 more than 65,000 Confederate troops camped in and around the town. The most devastating imprint of the war fell on July 30, 1864, when Confederate Gen. John McCausland levied a ransom of $100,000 in gold or $500,000 in greenbacks, or the town would be burned. The town refused to pay, so following breakfast on the square, he gave orders to torch the town. After the burning and looting were over, more than 500 structures lay in ruins. Forty-nine years later the Lincoln Highway marched through town, this time offering hope and prosperity.

Col. Benjamin Chambers, namesake and founder of Chambersburg., donated the land to build the first courthouse. More significantly, in 1767, he gave land to the Presbyterian, Lutheran, and German Reformed churches as an expression of his concern for religious tolerance. The only strings were annual payments to him or his descendants of a single red rose by each congregation. More than 200 years later, the "Red Rose Churches", as they are known, maintain the tradition in separate ceremonies.

Chambersburg is best known for giving birth to Pennsylvania's only United States president—James Buchanan. He was born April 23, 1791 at Cove Gap, a passage through the Allegheny Mountains southwest of Chambersburg. It would have been great to claim this pass as the route for the Lincoln Highway, but this was not to be. The Lincoln Highway negotiated the Alleghenies a few miles to the north.

Buchanan became the fifteenth President of the United States on March 4, 1857. A few historians rate his qualifications as second only to those of John Quincy Adams and Franklin D. Roosevelt. Perhaps his most significant presidential act was his prevention of the repeal of the Judiciary Act of 1789. This act would have given each state the right to interpret the constitutionality of state and the repeal of federal laws instead of the Supreme Court. The repeal would have meant the collapse of the Supreme Court and weakened federal laws.

"ORPHEUM THEATER IS BURNED, FIRE OF UNKNOWN ORIGIN COMPLETELY DESTROYS LEADING PLAYHOUSE..."

were the headlines of the Chambersburg *Public Opinion* on May 24, 1920. This is significant because on July 20 and 21, 1915 my father booked the Orpheum to show off the children and businesses of the community with his photographs. Just two months short of five years later the theatre was reduced to twisted steel and ashes. Good Will Fire Company records reveal the following:

RED ROSE CHURCH
The Presbyterian Church of Falling Spring

In 1767 Colonel Benjamin Chambers, the town's founder, donated land to three churches: Presbyterian, Lutheran, and German Reformed. His only request was that annually in June, each congregation give to him or his descendants a single red rose. The tradition is still maintained.

Chambersburg Free Library

Orpheum Theater

Orpheum was also known as Franklin Theater, which my father booked for July 21, 1915. Five years later, May 23, 1920, the theater burned to the ground from an unknown cause.

Chambersburg, Pennsylvania

Northwest corner of Lincoln Way and 3rd Streets.

Chambersberg Concrete Post

Northwest corner of Lincoln Way and Franklin Streets—one block east of Federal Street.

The fire gained much headway. Windows were framed with fire, and masses of smoke rolled from the eaves. The firemen, noticing that the fire was dense in the stage section, directed streams on that part of the building. At the same time it was noticed that the fire was raging along the wing galleries and an effort was made with poor streams to reach the flames.

Firemen broke into the stage freight entrance. This created a strong draft, and at 4:45 A.M. the upper portion of the building collapsed, falling into King Street. All efforts of saving the theater were abandoned, and attempts were made to save surrounding residences. The fire was marred by one accident. When the theater wall collapsed into King street, a member of the Good Will Company, was struck in the breast by a flying piece of tile. The missle tore a hole in his shirt and inflicted a minor cut in the flesh.

Loss: $33,000. Manager, H. R. Weber lost $2,000 plus business already contracted for.

4:15 A.M., May 23, 1920, the Orpheum Theater, West King Street.

From PUBLIC OPINION:

The Orpheum Theater had been opened October 12, 1911 by a vaudeville show presenting, "Madame Sherry". For reasons of space, cost and new building codes the theater was never rebuilt.

Big Novelty In Pictures

Chambersburg business men, business places and factories, Society men, women and children caught by the camera men on the streets. This all takes place on the screen at the Orpheum.

5 cents—ADMISSION—12 cents

Orpheum Theater ad in the *Public Opinion*, Chambersburg, Pa. Wednesday morning, July 21, 1915. This was part of the theater ad for my father's showing. In commentary, the *Public Opinion* said, "The thirty minutes of local scenes, local people and home industries was a popular treat to the Orpheum patrons yesterday matinee and night. This novel feature will be repeated today and for the last time. The shows will start at 2:30, 7:30 and 9:00."

ORPHEUM THEATER

EVERETT, *Pennsylvania*

IN HIS OWN WORDS:

Saturday, July 31

Fair & Rain. Run over to Everett, Pa. 16 miles. Booked and shot town for 3rd. Rained at night.

Tuesday, August 3

Rain. Run off slides of Everett, Pa. Show went $40.00. 50% 50

KARNS IS THE SURNAME of Alexander W. Karns, the Everett carriage builder, and his son, W. Chester Karns. The son became an early auto enthusiast and actually built his first auto in 1898, making all the parts by hand, but was not successful in persuading his father to help manufacture the machine, based on a carriage design.

However, Alexander Karns's antipathy did not dissuade young Chester and in 1904 he produced a small twelve-horsepower gasoline runabout with the help of five local boys: Jake Dunkle, George and Luther Grubb, Barney Foor, Sr. and Waldo Avey. This got his father's attention and they petitioned the federal government for a loan to build a car factory in Everett. Unfortunately, their petition coincided with a similar petition from an upstart in Detroit named Ford. Henry Ford got the loan and the rest is history. If fate had smiled in a different direction perhaps the Lincoln Highway could add a major auto manufacturing metropolis to its list of achievements. On the other hand, if that had happened other important Everett surnames like Barndollar and Sponsler may have been buried in an avalanche of big city stuff.

Everett began March 7, 1787 when John Musser from Lancaster, Pennsylvania, deeded 400 acres to Michael Barndollar (Bernthaler in German) of Frederick County, Maryland. In 1795 Michael Barndollar plotted out a town

Bloody Run Historical Society • Photo by Barbara J. Miller

Karn, 1904

The first Karn was in 1898—all parts made by hand. Only one copy was made in 1904. A federal loan for manufacture was denied effectively canceling future development. The one car was later stolen and now is owned by a Steve Barker from Indiana.

Barbara J. Miller

Karns Garage, early 1900's

Lincoln Highway (Main Street) in Everett —the birthing place and home of the Karn automobile

Barbara J. Miller

McDaniel Building, early 1900s

The rear of the building housed the theater my father booked for August 3, 1915 to show off the children and businesses he had photographed two days before. Notice the windmill on top of the building.

White settlers were attracted to this land for the same reasons the Shawnee and other Indian tribes inhabited the area. It was a good place to live—good fishing, close proximity to Mt. Dallas Gap, a break in the mountains, hills for game, and fertile fields for growing crops.

Everett today is a picture of activity and ambition. Nestled at the foot of Tussey Mountain, it includes excellent commercial opportunities, fine homes, active churches, modern schools and thriving businesses. Everett has proven to be a good place to live. Its citizens believe that the resources and capabilities have never been adequately tapped and are working on programs to further develop its natural resources.

A few other Everett surnames of history:

CULBERTSON, DUBOIS, GUMP, GIBBONEY, TUSSEY, FOOR, WIGFIELD, BENDER, ASHCOM, RITCHEY, LUCAS, SHEEDER, MYERS, MCCLURE, FELTON, HUGHES, STAILY, HENRY, and PADANRE.

on his 400 acres along Bloody Run Creek and called it Waynesburg in honor of Revolutionary War General "Mad" Anthony Wayne. Michael had eight children including son, Jacob, who became the first postmaster of the post office called – not Waynesburg – but Bloody Run. The name, "Bloody Run", dates to a ghastly massacre of forty-odd families on their way to Fort Bedford on August 5 and 6, 1763, near the present site of Everett. The small stream literally ran red with the blood of the victims, and was named by British soldiers. A court order 110 years later, in 1873, changed the name to Everett in honor of Edward Everett, former governor of Massachusetts, minister to Great Britain and a famous orator.

The Barndollar name has been permanently engraved in Everett history, popping up over the years in enterprises such as hotels, tanneries, blast furnaces, mercantiles, saw mills, lumberyards, taverns, and the present Everett United Methodist Church, formerly the Barndollar Methodist Church. The Methodist church blazed the church trail in Everett with its first building, in 1812. Following the Methodist example, the Lutheran, Presbyterian, Reformed, Baptist, and others later established congregations, all reflecting the religious tradition of Everett.

EDWARD EVERETT sat on the platform with Abraham Lincoln at Gettysburg and gave a stirring two hour speech followed by Lincoln's brief one. Both touched the same topics. Lincoln's was remembered.

Everett reportedly wrote Lincoln a note: "I should be glad if I could flatter myself that I came as near to the central idea of the occasion in two hours, as you did in two minutes."

Lincoln replied: "In our respective parts yesterday, you could not have been excused to make a short address, nor I a long one."

Grand Opening, May 2, 1896

My father may very well have shopped in this Red Front Store or even phtographed it for the theater show August 3, 1915.

GRAND OPENING
AT
BARNDOLLAR'S
RED FRONT STORE,
Main Street, Everett, Pa.
(NEXT DOOR TO M'CLURE'S HARDWARE STORE.)

Saturday, May 2, 1896.

A COMPLETE LINE OF
DRY GOODS, NOTIONS, SHOES, GROCERIES, FLOUR, FEED, GLASSWARE, &c.

This entire stock has been purchased by J. J. Barndollar, who has had many years' experience in buying for this community, and knows the wants of the people in this section. The stock of goods will be sold at the very lowest prices. If you will visit our store we will convince you that it will pay you to buy from us. Our expenses are light and our stock having been purchased at ROCK BOTTOM PRICES we feel assured we can save you money. Bear in mind the fact that our entire stock is new. No old or shelf-worn goods in our store. Hoping to receive a share of the public patronage, we are

Very respectfully,

J. J. BARNDOLLAR,
Manager BARNDOLLAR'S RED FRONT STORE.

☞ We will quote prices in the newspapers next week.

Republican print, Everett, Pa.

Barbara J. Miller

Barbara J. Miller

Barndollar Methodist Church, 1910

Jacob Barndollar, son of Michael Barndollar, founder of Everett, built this church 1859-1860, now called The Everett United Methodist Church.

Everett, Pennsylvania

Lincoln Highway Heritage Corridor sign two miles east of Everett. Note the Allegheny Mountains.

Everett, Pennsylvania

Northwest corner of Hillside Street and Route 30, just before the curve to the right.

Barbara J. Miller

Lincoln Highway, early 1900s

Looking east on Main Street. McDaniel building on the left also housed the theater. Note the dirt street lined with trees. Later paving proved incompatible with trees.

Barbara J. Miller

Lincoln Highway, early 1900s

Just east of Everett is the Lincoln Highway my father traveled in 1915, just before arriving in Everett. The right center fields are now the Down River Golf Course.

BEDFORD, *Pennsylvania*

IN HIS OWN WORDS:

Thursday, July 29, 1915

Cloudy. Left 6 o'clock for Bedford, Pa. Arrived at 4:30 P.M. Booked for Monday 2nd. At Dreamland Theater. Rough traveling.

Friday, July 30, 1915

Fair. Shot up Bedford, [not a good choice of words eighty-five years later] *for show Monday night. Flashes and kids* [in 1915 photos were taken by a "flash," igniting a small amount of gun powder while squeezing the camera bulb].

Sunday, August 1, 1915

Rain and cloudy. Loafing all day in camp at end of North Street, Bedford, Pa.

Monday, August 2, 1915

Rain. Run slides for show and proofs (Dreamland Theater). Show went $30.95. 50%—50%.

Thursday, August 5, 1915

Fair. Left Bedford at 6 A.M. Arrived in Beaver, Pa. 140 miles at 10:30. Good roads, fine weather. No trouble.

Dreamland Theater Building, 2002

Downtown on Pitt Street (Lincoln Highway) in Bedford. My father booked this theater August 2, 1915 to show off the photographs he took of local children and businesses. The building now houses Gardners Candies.

IT ALL STARTED AS AN INDIAN PATH, the Lincoln Highway in Bedford, that is. In the early 1700s a maze of Indian trails in western Pennsylvania wound their way through virgin forests into an east-west trade route used by early settlers to what ultimately became Pittsburgh. In 1750 a Scottish adventurer, Robert MacRay, established the first recorded trading post along a branch of the Juniata River to trade with Indians, settlers and trappers. During the next 10 years, by 1760, General John Forbes built a "road" (little more than a

The Lincoln Highway over the years was a critical link between the eastern cities and Pittsburgh to the west, and to the pioneer vistas beyond. Inns, stores, miscellaneous businesses and courthouses sprouted along its stem. Bedford itself boasts important places like Espey House, where George Washington bivouacked in 1794 during the Whisky Rebellion; the oldest Pennsylvania courthouse still in use, and the famous curative springs at the Bedford Springs Hotel. President James Buchanan used this hotel

Henry Kline Siebeneck

General John Forbes, 1750's

This is the man whose vision and determination turned Indian paths into what became the Lincoln Highway and ultimately set the stage for the country's first Interstate Highway—the Pennsylvania Turnpike in 1940

walking path) from Carlisle, Pennsylvania, past the site of MacRay's trading post (Rays Town) and on to Fort Pitt. Transportation of supplies to Fort Pitt was important to the British so they hired area settlers to widen the path into a road able to accommodate freight wagons. It proved important not only to the military interests of the British in their fight against the French and Indians, but also to the area economy. The road was widened again and again until, in 1913, part of it officially became the Lincoln Highway. In 1758 Fort Bedford (named for the Duke of Bedford) became Bedford.

Thus was born "Mother Bedford." Lands obtained from the Indians from 1754 to 1768 and again in 1771 became Bedford County, encompassing most of western Pennsylvania. Out of this Bedford County were born the present-day counties of Huntington, Fulton, Blair, Cambria, Somerset, Westmoreland, Fayette, Washington, and Greene along with portions of Center, Indiana, Armstrong, Allegheny, and Beaver.

Pioneer Historical Society

Forbes Road, 1750s

A remnant of the old Forbes Road. Beginning as an Indian trail the Forbes Road became a strategic element of the British French and Indian War. The trail was expanded from time to time, portions of it eventually becoming the Lincoln Highway in 1913.

Founders Crossing Building, 2002

A building often passed by my father in 1915. Today it houses two floors and a basement of antiques, gifts and memorabilia, plus an informal setting for lunch. Located on the southwest corner of Pitt Street and Juliana in Bedford.

Kevin Kutz

Kutz's Dreamland Theatre

Kevin Kutz, local artist, spent two years (1981-82) painting this picture of Pitt Street (Lincoln Highway) from the roof of a building across the street. The theater is the second from the left with the fluted roof.

Fort Bedford, 2002

Fort Bedford was built by the British as part of the French and Indian War. It is now the Fort Bedford Museum. The Juniata River is on the left

for his summer White House, performing one of the most memorable acts of his presidency by receiving the first trans-Atlantic telegram from Queen Victoria in England. The hotel was also host to Presidents Rutherford Hayes, James Garfield and Benjamin Harrison, plus justices of the United States Supreme Court. It was the only time the justices met outside their Washington, D.C., chambers. Historians speculate that the Dred Scott case was debated on the hotel porches. The controversial 1857 decision permitted slave owners to pursue and capture runaway slaves in free states and territories. Maybe they would have done better if they had stayed home.

Modern Bedford is home to more than 3,000 people, the Lincoln Highway, and three Interstates: I-76, I-70, and I-99.

Administrator Kay Williams and Joe Riggles from the Pioneer Historical Society deserve special mention for going out of their way to provide a delightful view of Bedford's history. Joe also introduced my wife and me to Kevin Kutz, a local artist, who painted a 20-foot long mural of downtown Bedford on the inside north wall of the Founders Crossing building. The mural included the Dreamland Theater. Kevin later looked us up in the Founders Crossing building while we were having lunch and gave us a signed print of the Dreamland Theater that he painted in 1981-82 from the roof of a building across the street.

Coffee Anyone?

The famous Lincoln Highway Coffee Pot Café is shown in 2002 in sad disrepair. Happily the Licoln Highway Heritage Corridor obatained a grant to remove the pot. We now can reorder that cup of coffee. The structure has been moved to a park across the Lincoln Highway.

Bedford Concrete Post

East of the intersection of Routes 30 and 31 on north side, just east of Jean Bonnet Tavern

Bedford Courthouse

The oldest continuously functioning courthouse in Pennsylvania. It was built in 1828.

An Ignoble End, 2002

On Friday morning, October 26, 2001, a longstanding Lincoln Highway landmark, the Grand View Point Hotel, burned to the ground. All that remains is a grotesque view of twisted steel beams. The great ship hosted such famous guests as Clara Bow, George Burns, Joan Crawford, and Greta Garbo. Its view was legendary where one could see into three states and seven counties. It was located 17 miles west of Bedford, in Bedford County.

Seven Mile Stretch looking west

Seven Mile Stretch looking east

Heading west just over the Summit County line at Bald Knob summit, the "Seven Mile Stretch" strings out ahead to the horizon.

Burma Shave

BURMA SHAVE TRAVELED THE ROAD of ingenuity, failure, ridicule, old-fashioned kindness, relative enterprise, poetic license, and accident before it found itself on little signs along the road outside Minneapolis, Minnesota. Grandfather Clinton Odell accepted an offer of an unemployed chemist in return for an earlier $25 gift toward the chemist's medical bills.

They had experimented with nearly 300 different shaving cream formulas that didn't work when by-guess-and-by-golly, Clinton discovered that the three-month-old #143 formula gave a good shave, given a few months to cure. Thus it was that Burma Shave, against professional advertising advice, found itself on a series of six small red signs with white letters ten to twenty yards apart along two lane roads all across the country in the 1930s and 1940s. Interstates, wide rights-of-way, and increased speed limits ushered the humorous limericks into the roadside hall-of-fame (If there is none, there should be).

Here are a few examples:

DON'T LOSE YOUR HEAD	DROVE TOO LONG	BROTHER SPEEDER	SPEED WAS HIGH
TO GAIN A MINUTE	DRIVER SNOOZING	LET'S REHEARSE	WEATHER WAS NOT
YOU NEED YOUR HEAD	WHAT HAPPENED NEXT	ALL TOGETHER	TIRES WERE THIN
YOUR BRAINS ARE IN IT	IS NOT AMUSING	GOOD MORNING NURSE	X MARKS THE SPOT
Burma-Shave	Burma-Shave	Burma-Shave	Burma-Shave
THE MIDNIGHT RIDE	THE ONE WHO DRIVES	PASSING SCHOOL ZONE	AT INTERSECTIONS
OF PAUL FOR BEER	WHEN HE'S BEEN DRINKING	TAKE IT SLOW	LOOK EACH WAY
LED TO A WARMER	DEPENDS ON YOU	LET OUR LITTLE	A HARP SOUNDS NICE
HEMISPHERE	TO DO HIS THINKING	SHAVERS GROW	BUT IT'S HARD TO PLAY
Burma-Shave	Burma-Shave	Burma-Shave	Burma-Shave

WEST VIRGINIA
And The Lincoln Highway

Miles: 5 Towns: 1

IN MARCH, 1928, A CHANGE WAS MADE in the route of the Lincoln Highway from Pennsylvania to Ohio, which avoided a congested route along the north side of the Ohio River in Pennsylvania. The new route passed through the towns of Crafton, Imperial and Clinton in Pennsylvania and Chester in West Virginia. Thus the Mountain State became the 13th state just in time to be monumented by concrete posts in September of that same year.

Three of these posts are in Chester but we could find only two. We were told there used to be one outside a convenience store, but that place of business is long gone and we could find no post at the site. This is a unique section of the Lincoln Highway where one can travel in three different states in just seven miles. Unfortunately for my father this change came 13 years too late. He took the "congested" route passing into Ohio north of the river just west of Glasgow.

Chester Teapot, 2002

Bedford, Pennsylvania has the famous Lincoln Highway Coffee Pot, but Chester, West Virginia, has the world's largest Lincoln Highway Tea Pot.

Chester, West Virginia,

West side of California Ave. (Old Lincoln Highway), north of First Street

Chester, West Virginia

Northeast corner of Virginia Ave. and 3rd Street

Ohio River Bridge

The Route 30 bridge over the Ohio River at Chester, West Virginia.

Route 30 Bridge

The Ohio River Bridge at Chester, West Virginia as seen from the Lincoln Highway remnant at the Ohio River Overlook.

Lincoln Highway Remnant

The Ohio River Overlook is all that remains of the former Lincoln Highway Bridge.

Autocar Trucks 1897

Time Life History of the United States, Vol. 10

A section of the Lincoln Highway is being marked by workmen operating from the two Autocar trucks donated by the Autocar Company through David S. Ludlum, its president and a founder of the Lincoln Highway Association. Note that the post being painted is probably wood and predates the concrete posts of 1928.

AUTOCAR IS THE OLDEST VEHICLE nameplate in continuous production in the United States. The company was founded in 1897 by the brothers Lewis and John Clark as the Pittsburgh Motor Car Company. Its first vehicle was a gas-powered tricycle. In 1899 the company moved to Ardmore, Pennsylvania, on what would become the Lincoln Highway, and renamed the firm Autocar which became a registered trademark in 1905. The company began its focus on trucks in 1907 and introduced its first commercial vehicle in 1908 with continued manufacture to the present.

In 1953 the controlling interest in the company was purchased by the White Motor Company and moved twenty miles west on the Lincoln Highway to Exton, Pennsylvania.

In 1981 Volvo purchased White, renaming the company, Volvo White Truck Corporation, retaining the Autocar name for heavy construction work. In 1997 Autocar celebrated its 100th anniversary.

In July, 2001, the Grand Vehicle Works (GVW) of Highland Park, Illinois acquired Autocar from Volvo creating a new subsidiary, Autocar LLC and thus positioning it for another 100 years of successful contributions to the trucking industry.

OHIO
And The Lincoln Highway

264 miles 33 towns

THE OFFICIAL LINCOLN HIGHWAY ROUTE through Ohio was for many years a work in progress. It involved as much or more controversy than in any other state. Major game players were the towns of East Liverpool, Lisbon, Canton, Wooster, Ashland, Bucyrus, Upper Sandusky, Ada, Beaverdam, Lima, Delphos, and the U.S. Army convoy in 1919. This drama, for the most part, was played out after my father "soared" through the Buckeye State in only thirteen days in 1915 which, please note, included ten days spent taking photographs in Minerva. Of the thirty-three Lincoln Highway towns in Ohio, my father records only four: Lisbon, Minerva, Wooster, and Ada. His main interest was Minerva, where he photographed children and businesses, rented a theater to show off the children's pictures and at the same time built his travel chest for the rest of the trip west. The other towns were merely a means to an end—San Francisco.

The final choice for the Lincoln Highway in Ohio was accelerated by the federal government in 1925 with its decision to replace named highways with numbers. The Lincoln Highway became Federal Highway Route 30, taking over, for the most part, Ohio state route #5. This transition was a long process with even the 30N and 30S compromise not fully resolved until 1973.

In 1928 the Lincoln Highway Association arranged for the "planting" of nearly 3,000 concrete marker posts from New York to San Francisco. Of these, 241 earned sentinel spots in Ohio with possibly two dozen still remaining, only three of these are thought to be at or near their original locations, having traded their functional role for a monument in history.

When it comes to auto production, Ohio's record is not all that shabby, since 545 different makes were produced within the state with 30 of them coming from Lincoln Highway towns. It would take a very dedicated autophile to recognize any of the names today, much less to expect that any would survive to grace the modern Lincoln Highway. They wore strange names like Hydrocar, Leach, Worldmobile and Shunk.

Brick Remnent in Summer Atire 2001

Baywood Street between Minerva and Robertsville in June.

Brick Remnent in Winter Garb 2002

A light January snow covers the brick on Baywood Street between Minerva and Robertsville. The brick, no doubt, came from nearby Malvern, Ohio, the "Paving-Brick Capital of the United States."

Ohio *Pillars and Posts*

THE LINCOLN HIGHWAY HAS A UNIQUE RELATIONSHIP with the State of Ohio—not only in the convoluted choice of its path but also in its memorabilia. In addition to at least eight 1928 style concrete posts, Ohio can boast of at least five brick pillars and one Memorial.

PILLARS

Hanoverton, Ohio

Lincoln Highway pillar, NE corner of highway 9 and Lincoln Highway U.S. 30.

Bucyrus, Ohio

Round Lincoln Highway stone pillar, 1/4 block east of Whetstone Road on north side of Lincoln Highway U.S. 30.

Crestline, Ohio

Lincoln Highway pillar dedicated to A. F. Bement, VP and secretary of Lincoln Highway Association, May 1, 1922. SW corner of Lincoln Highway U.S. 30 and Clink Blvd.

Wyandot County, Ohio

L. A. Kuenzli farm, five miles west of county line on south side of the Lincoln Highway U.S. 30 between Upper Sandusky and Bucyrus. Identical pillars mark each side of the driveway entrance.

Special Collection,
University of Michigan Library

Crestline, Ohio

Lincoln Highway pillar dedicated to J. F. McMahon, first Lincoln Highway Crestline consul, May 1, 1922. SE corner of Lincoln Highway U.S. 30 and Clink Blvd.

Osceola. Ohio

Replacement piillar set summer, 2001, by LHA Ohio Chapter.

Special Collection,
University of Michigan Library

POSTS

Hanoverton, Ohio

One block from Spread Eagle Tavern on the Lincoln Highway U.S. 30.

Dalton, Ohio

Looking pretty good, in a private driveway on the south side of the Lincoln Highway U.S. 30, west part of town.

Minerva, Ohio

SE corner of Market Street and Lincoln Highway U.S. 30.

Moscow, Ohio

South side of old Lincoln Highway, 1/10 mile west of Highway 52 east of Riceland, Ohio.

East Canton, Ohio

Just west of Cedar Street on the Lincoln Highway U.S. 30.

Riceland, Ohio

Northeast corner of Highway 57 to Orrville, and Lincoln Highway U.S. 30 under a flowering crabapple tree.

Dalton, Ohio

At the First National Bank, corner of Mill Street and Lincoln Highway U.S. 30.

Van Wert

West of Jefferson Street, south side, Van Wert, Ohio.

Bucyrus, Ohio

Memorial to John Hopely, champion for "good roads" everywhere. He was an enthusiastic promoter of the Lincoln Highway and was Ohio state consul from the earliest years until his death in 1927. The monument was built in 1929 along the Lincoln Way on the grounds of the Bucyrus Country Club. It was built of stone—special rocks from places connected with John's life such as Elkton, Ky., where he was born; Southampton, England, and Montevideo, Uruguay, where he served as U.S. consul; and one was from Lincoln's birthplace.

Mifflin, Ohio

South side of Lincoln Highway U.S. 30, 1/2 block west of Highway 603

Mifflin, Ohio

Lincoln Highway loyalty is alive and well in Mifflin.

Spread Eagle Tavern

The Spread Eagle Tavern, built in 1837 by Will Rhoades, is only one block from the Lincoln Highway in Hanoverton, Ohio. It was built to serve the prosperous business community created by the Sandy and Beaver Canal, and over the years has become a famous landmark.

1915
ANNUAL COST OF LICENSE TAG

New York	$35.40
New Jersey	40.50
Pennsylvania	40.50
Ohio	25.35
Iowa	30.35
Massachusets	40.50
Michigan	35.40
Minnesota	30.35
North Dakota	30.35
California	75.80

1924
SUGGESTED CAR EQUIPMENT FOR THE TRANSCONTINENTAL MOTORIST

Car Equipment

1 Lincoln Highway Sustaining Membership Radiator Emblem
1 pair Lincoln Highway pennants
2 sets of tire chains
6 extra Cross Chains (for above)
1 set chain tightener springs
1 set tools
2 jacks
1 pair good cutting pliers
2 extra tire casings, mounted
1 casing patch
3 spark plugs
8 feet of high tension cable
1 gallon can of oil
1 hand axe
1 shovel (medium size)
1 upper radiator connection
1 lower radiator connection
1 set lamp bulbs
1 flash light
3 extra fuses
1 tow line
1 auto–robe trunk

Camping Equipment

1 5-gal milk can
1 canteen, 2 quarts
1 frying pan, 10 inch
1 grate for camp fire, 12X24 inches
1 coffee pot, 2 quarts
4 cups, large
4 pans (deep) 5 inch diameter
4 knives
1 auto camp stove
6 forks
6 teaspoons
2 cooking spoons
4 soup spoons
1 dipper
8 plates, 8-inch diameter
2 stew pots (to nest)
1 cooking fork, 3-prong
1 carving knife, butcher type
3 bars Ivory soap
6 dish towels
1 can opener
1 bread pan
1 bucket with lid
1 can for pepper
1 patent egg carrier
1 cork screw
1 air-tight coffee can
1 air-tight tea can

Update: All one needs today is a cell phone, credit card, and sunglasses.

51

LISBON, *Ohio*

IN HIS OWN WORDS:

Friday, August 6, 1915

*Fair. Arrived at Lisbon, Ohio at 5 P.M..
97 miles [from Beaver, Pa] Bad dirt roads,
ditched 4 miles out of East Liverpool, Ohio.*

Saturday, August 7, 1915.

Fair. Left Lisbon at 7 A.M. for Minerva.

THERE ARE FEW MODERN CIRCUMSTANCES to help one feel with the pioneers of 200 years ago. Did they feel deprived without running water, the ability to speak instantly to someone ten miles away—or half-way around the world—to see their images as seen by others as in a photograph, to fly like an eagle, or to travel to Grandma's house in an air-conditioned capsule in August? They lived within the boundaries of the world as it was. "We never knew we were poor" was the reflection of Sam Levinson in *Everything But Money*.

But this is the stuff from which our country was made, and so it was for Lewis Kinney, first official property owner of Lisbon, Ohio in 1802. He later bought Section 14 of Center Township, Columbiana County, Ohio, on August 7, 1805 by patent deed from the United States signed by President Thomas Jefferson and James Madison. As recently as twenty years before the deed was signed, Lisbon was a wilderness of trees, streams, and wild animals, occupied by Indians. Before the creation of the United States as a government, Lisbon was populated by German, Lutheran, and British and Scotch Presbyterian families from Pennsylvania, Virginia, and Maryland.

Between 1803 and 1805 Lewis Kinney built the Old Stone Tavern, the earliest stone structure to survive in Ohio. It is now the Old Stone House Museum restored in 1951 by the Lisbon Historical Society. Lewis Kinney also served as Ohio State Senator from 1808 to 1813.

The first newspaper was *Der Patriot Am Ohio* published in German by William D. Lepper in 1808, in a log cabin on North Beaver Street. It was later published in English on the Lisbon Square and renamed, *The Ohio Patriot*.

A significant bit of history is the use of brick in housing construction. The State of Ohio has become famous for its brick. In Lisbon, the Picking Store, built in 1806, is the oldest brick building in Ohio, and brick dominates the center of town, the square and nearby

Lepper Library – 1897, 2002

The Lepper Library, on the Lincoln Highway, was donated to the community by Virginia Cornwell Lepper in memory of her husband, Charles W. Lepper. Since the dedication, May 24, 1898, it has served the literary needs of every generation. The building was designed to emulate an old English Country Church of the early 1800s. It is the most striking building in Lisbon.

Old Stone House Museum, 1803, 2002

The oldest stone house in Ohio, was built by Lisbon founder, Lewis Kinney as a tavern. It is now the Old Stone House Museum.

streets. In fact, Ohio brick is responsible for some of the best preserved brick sections of the Lincoln Highway. Near the close of the nineteenth century, Lisbon's increased prosperity was due largely to its status as the county seat and the growth of industries such as coal and the Thomas China Company. This increased prosperity expanded housing throughout the town and resulted in different architectural styles surrounding the older brick of central Lisbon. The contrast is easily observed when approaching the square on the Lincoln Highway from either east or west.

Modern Lisbon is a rural village of 3,037 (1990 census) and the county seat for Columbiana County. Agriculture, coal, mills, tanneries, and pottery still dominate the economy of Lisbon. In recent years the focus is on metal fabricating, hoist manufacturing, coal and agriculture.

A visit to Lisbon is a rewarding experience for anyone who enjoys the Lincoln Highway.

Wirtz Building—1808, 2002

The first brick hotel in Lisbon. My father may have stayed here August 6, 1915.

MINERVA, *Ohio*

IN HIS OWN WORDS:

Saturday, August 7, 1915

*Arrived in Minerva [from Lisbon] at 9 a,m, 18 miles.
Bad dirt roads. Ging [an associate] got sick.
Left for home.*

Sunday August 8, 1915

*Rain. In camp at Fair Grounds, Minerva, O.
Booked Minerva, O for 11,12,13th at Dreamland Theater.*

Monday, August 9, 1915

Fair. Made kids shots and flashed for show 11th.

Tuesday, August 10, 1915

*Fair. Made kids shots in Minerva, E. Rochester
and Bayard for show in Minerva.*

Wednesday, August 11, 1915

*Rain. Run slides and advertised for show.
Receipts $38.44. 50% 50%*

Thursday, August 12, 1915

*Fair. Made shots in Malvern for show in Minerva Friday.
Receipts this night $20.90*

Friday, August 13,1915

*Fair. Passed proofs in Minerva and showed.
Receipts, $29.00*

Saturday, August 14, 1915

*Fair. Delivered some in Minerva and passed proofs
in Malvern.*

Sunday, August 15, 1915

*Fair. Run off Minerva and Malvern. In camp all day
(Minerva).*

Monday, August 16, 1915

*Fair. Delivered in Minerva & Malvern
Left 12:00 noon*

THE HISTORY OF MINERVA, OHIO, can be understood best in the context of Stark County. The area now known as Ohio was originally occupied by the Erie, Huron, Ottawa, Tuscarora, Mingo, Delaware, Shawnee, and Miami Indians. Between 1669 and 1670 it was visited by Frenchmen like Robert Cavalier, sieur de La Salle who extensively explored and mapped the area. This led to France's claim to the entire Ohio valley, and triggered the French-Indian War with the British. The creation of the Northwest Territory by the Continental Congress on July 13, 1787, and the Land Ordinance of 1785, authorized the sale of one-mile square sections of land and led to the development of Ohio.

On February 13, 1808, by an act of the state legislature, Stark County was drawn from land originally in Columbiana County. It was officially organized January 1, 1809. Osnaburg was one of the eight townships making up the county of Stark prior to 1815.

ENTER MINERVA

In retrospect, July 5, 1833, was an important day for Minerva. It was on this date that John Whitaker, a Quaker land surveyor, went to the county courthouse in Canton, Ohio and recorded the 123 acres he had purchased twenty years earlier in 1813 from Osnaburg Township. Through this land flowed the Little Sandy Creek, and on its banks John Whitaker built the first gristmill in 1832.

Within the next ten years, John Whitaker's mill became the magnet for other pioneers who put down their roots while putting up their cluster of log cabins. One of the first millers was Pim Taylor, a teamster who operated two wagons between Philadelphia and Mansfield, Ohio. He decided to settle down in Minerva and married Keziah Whitaker, John Whitaker's niece, whom he no doubt met in his travels through Minerva.

Dreamland Theater

My father booked this theater August 11, 12, and 13, 1915, to show off the Minerva children and businesses he photographed. Today it is home to the Berea Bible Church.

Reducing the history of peoples' lives to a few paragraphs nearly 200 years after the fact does not do them justice, nor even comes close to understanding the stage upon which they played out their lives. For instance, how did they treat the common cold, or worse, an infected appendix? About what did they laugh? What did they do for entertainment or how did they get the news without TV? What did they do with leftovers without a refrigerator? How did they enjoy a bath without a shower? How did they appreciate each other without deodorant? What with carrying all their water, hand scrubbing their clothes on a washboard, preparing every meal from scratch, when was there time to enjoy a sunset or the sound of rain on the roof or to just sit back and dream? But we know there were moments of joy and celebration because modernity has no monopoly on laughter and a good time. One of those joyful moments must have been the birth of Pim and Keziah Taylor's first child, named Minerva Ann. In fact, the community was so won over by this child's charm, they named their village in her honor, "Minerva." She was born July 14, 1828 and died 87 years later, just a few months before my father came to town on his way to San Francisco in 1915—on the Lincoln Highway.

Do you wonder why the people of Minerva would so honor this girl in a time when society did not easily grant great prominence to females? Why not "Taylorville" in honor of Pim Taylor, or "Johnstown" in honor of John Whitaker, or "Sandy Creek?" The answer may be buried in dusty files or memoirs somewhere, but the chances are we will never know.

However, it's not the first time little girls' names have risen to the high status of household words. It might be appropriate to mention "Mercedes" since we are in an automotive frame of mind. It was in 1901 the

Minerva Area Historical Society

This gas station *still stands in downtown Minerva where it began about 1910.*

Minerva's first non-wood bank building.

Through the years it has been a grocery store and meat market. Today it is a dental office for Daniel H. Thomas Jr., DDS.

German Daimler Company raised the bar for automotive style, mechanical innovation and speed, with its new automotive creation. Under the direction of Daimler's engineering assistant, Wilhelm Maybach, and at the urging of a wealthy client, Emile Jellinek, for whom the car was built, the new car was an immediate success and well received with some of its features still in use today. Strangely though, it was not called "Daimler." It was called "Mercedes," after Emile Jellinek's oldest daughter. Now you know.

Another significant industry in Minerva began when the Pinnock brothers built a foundry to manufacture plows and other items used by the early settlers. In this factory, along the early Pennsylvania Railroad, the first steel railway car was conceived and manufactured. William Pennock received his U.S. patent in 1880. Some years later, Minerva lost the railroad car business to a wealthy industrialist who bought the patents and moved the operation out of Minerva.

Today, Minerva is a neat rural Ohio town of around 4,000 people, and the largest between East Liverpool and Canton.

Minerva Post

Southeast corner of Market Street and the Lincoln Highway.

MINERVA, *The Lost Gold Village*

THE YEAR IS 1755 AND A FRENCH CONVOY has stopped in the Ohio wilderness during the French and Indian War to unload a huge supply of gold and silver bars. The bars they are burying are worth approximately four million dollars, and are being removed from Fort Duquesne because of France's ongoing war with Britain.

The convoy that buried the gold and silver consisted of ten men and sixteen pack horses. When the men were ordered to stop and bury the treasure, they had no idea that the battle at Fort Duquesne had ceased. They covered the buried treasure with leaves, and hid their shovels under a log on a hillside. Suddenly the British began firing upon the convoy and eight of the ten men were killed. The two men whose lives had been spared jammed a rock in the fork of a tree one-half mile from the buried gold and silver, and they also carved a deer into a tree one mile away. They claimed that the treasure was buried in the center of a square created by four springs.

The treasure is believed to be located six to seven miles east of Minerva. Many of the signs mentioned above have been found, such as the shovels, the odd-shaped stone in the fork of the tree, the deer carving, and skeletal remains.

There have been many attempts in the past to find the bullion, but all have been unsuccessful. Prospectors have used witching rods, goldometers, and many other types of metal detectors, but their only discovery has been heaps of decayed rubbish.

One can only speculate on its fate. The area could have changed geographically, some wandering stranger may have stumbled upon it, or the tenth member of the convoy (who was never accounted for) may have returned for it. Perhaps the treasure is still buried.

Selected from an article by Carol Nolling in *Countryside Magazine*, 1989

WOOSTER, *Ohio*

IN HIS OWN WORDS

Monday, August 16, 1915

Arrived 8 miles west of Wooster. 60 miles. Good roads.

Tuesday, August 17, 1915

Rain. Yesterday seen auto smashed at RR crossing in Massillon, O.

LONG BEFORE THE WHITE MAN CAME (1700-1750), the Delaware and Wyandot Indians were enjoying the riches and natural beauty of the Wooster area.

The city of Wooster began with a single log cabin in 1808 along with the first wagon road through dense timber from Wooster to Massillon—undoubtedly ancestor to the Lincoln Highway, born 103 years later.

The federal census of 1810 lists a population of 332 in Killbuck Township (with an increase to 1,935 by 1820), made up of pioneers from Maryland, Virginia, New Jersey, New England, and Pennsylvania. The Pennsylvanians brought with them a dialect of the German language known as *Pennsylvania Dutch.*

The first Amish settler, Jacob Yoder, in 1817 began a migration of Amish, making Wayne and Holmes counties the largest Amish settlements in the United States.

Despite local industrial growth, the Wooster community retains its enviable agricultural image, with roots going all the way back to 1809, when Joseph Larwill planted his first corn.

Education also planted its roots early when Carlos Mather, the first real schoolteacher, arrived from Yale in 1818. In 1865 an itinerant Presbyterian minister stopped on a hill north of town and was so impressed by the view that he lobbied the Ohio Synod of Presbyterian churches to establish a college there. These roots ultimately produced the College of Wooster, incorporated in 1866 and opened for classes in 1870 with 30 students —tuition, $30, room and board, $1 per week.

Lest the outsider be deceived by the fertile soil, neat rows of corn and soy beans, and meticulously cared-for barns and houses, this community has still other roots, such as oil and gas. In 1815 Joseph Eicher opened a salt works on Killbuck Creek, where one day a worker started a huge fire by dropping a hot coal into oil which had come up from the ground,

Lincoln Highway Remnent, June 4, 2001

Ten-foot wide brick remnant with concrete border, (Sylvan Rd., Wooster) – located one mile east of the intersection of Market and Liberty Streets. Go east on Pittsburgh Street to Bauer Rd. then south one block to Sylvan Rd. The brick remnant dead-ends at Apple Creek. The road was not brick in 1915 but it is the road my father traveled into Wooster.

Wayne County Courthouse

Centennial Celebration, August 11-15, 1896, honoring 1796 when Gen. Arthur St. Clair proclaimed "Wayne County."

If the Delaware Indian Chief, Beaver Hat, who had a cabin and an apple orchard at the site 200 years ago, should return today—just the thought challenges the imagination. Even my father's imagination would be stretched to see Wooster in its post-1915 dress. New bridges, highways, bypasses, industry, housing developments and shopping malls have all rearranged the face of Wooster. Only a small piece of one-lane brick Lincoln Highway remains on Sylvan Road just south of Pittsburgh Street as it crosses Apple Creek.

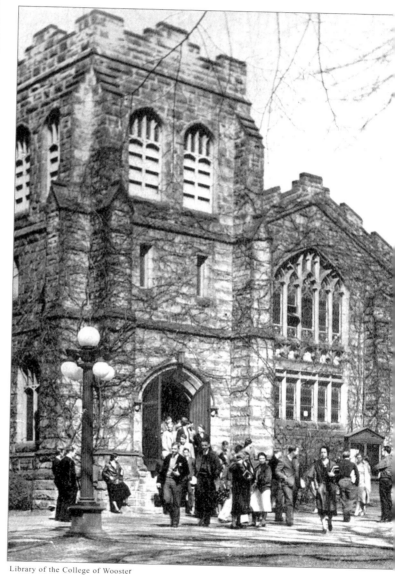

Memorial Chapel, College of Wooster

Built in 1901, my father could have seen it in 1915. This building was torn down in June, 1969.

instead of salt water. This was considered a disaster. Later it would be recognized as the means of lubricating and propelling those horseless carriages over the Lincoln Highway.

Wooster moved up to a second-class city in 1868 and by 1870 had a population of 5,419. In 1878 authorization was granted to build a $75,000 courthouse, which building has retained that honor to the present, or maybe only half an honor since the original plan was never finished leaving only a "half" a courthouse. In any case my father likely saw it when he drove through Wooster on August 16, 1915.

The 1,935 Wooster pioneers of 1820 had increased to 8,204 by 1920. In 1932 Mrs. Herbert Hoover, wife of the president, was guest of honor at the city's 125th anniversary. The manufacturing giant, Rubbermaid, Inc. was born in Wooster in 1934 when James R. Caldwell made a dustpan.

PACKARD Ask The Man Who Owns One

THE PACKARD WAS NOT ONE OF THOSE machines born in a basement on a shoestring. Packard emerged from an educated family with manufacturing experience, money, the will to succeed and a factory—the New York and Ohio Company, manufacturing incandescent lamps and transformers in Warren, Ohio. From the very beginning, quality was a requirement which paid off in both the short and long run. "Packard" always had a good ring to it, if not from experience, then from association, or observation or "I wish." It was always an uptown car—several cuts above its nearest competition. Even its conception and birth were upscale.

Brothers J. W. and W. D. Packard had a keen businessman father who made a fortune on Erie Railroad construction. He lived in a mansion on Main Street in Warren, Ohio, and had sufficient financial resources and a pool of mechanical genes from which to draw. James Ward Packard was two years younger than his brother, William Daud. He was born November 5, 1863, in Warren, Ohio, entered Lehigh University at age 17 in 1880 and graduated with a degree in mechanical engineering in 1884. On June 5, 1890, he and his brother, William, founded the Packard Electric company and in 1892 the New York and Ohio Company.

The Packard Club

Ward's submerged interest in horseless carriages surfaced with the first issue of the *Horseless Carriage Magazine* in December 1895. In July, 1898, he made a $1,000 deposit for the thirteenth of twenty-two Winton cars built that year. Precise history here is vague with a touch of legend. It is accurate to say, though, that Ward picked up his new Winton in Cleveland on August 13, 1898, and drove all day (50 miles) to within four miles of his home in Warren. The final four miles of the journey was made behind a horse after dark. Here the story gets murky. One version has it that Ward was so upset with his Winton that he complained directly to Alexander Winton and received a reply to the effect that—"the Winton wagon as it stood was the ripened and perfected product of many years of lofty thought, aided by mechanical skill of the highest grade, and could not be improved in any detail, and that if he, Packard, wanted any of his own cats or dogs worked into a wagon, he had better build it himself as he, Winton, would not stultify himself by any departure whatever from his own incontestably superior productions." This may be one of those "true stories" that never happened. If the conversational reality between these two great auto pioneers has not come to light by now, it probably never will. We do know, however, that on April 6, 1899 the first Packard was built with the assistance of two former Winton employees, George Weiss and A. C. Nelson., with improvements and innovations over the Winton. Packard was the first to abandon the tiller for a steering wheel (1901), and use the automatic spark advance. We know, however, that Packard was not pristine. Witness the 1903 model K which was so bad that Packard recalled for recycling all but one of the thirty-four made. In that same year, 1903, the Packard plant was moved to a new sixty-six-acre plant in Detroit.

First Packard, Old #1, November 6, 1899, currently on display at Lehigh University, Bethlehem, Pennsylvania.

National Packard Museum

Packard from the very beginning enjoyed a reputation for fine quality. Some said it was over engineered. Packard's demise in 1956 was claimed by some loyalists to be partly Packard's own fault because the cars were made so good that owners just kept them instead of buying new ones. Among early well known Packard owners was William D. Rockefeller, who moved from Winton to Packard with the purchase of two at America's first auto show in New York in 1900.

Packard was also one of the first auto companies with a slogan that stuck. In response to a request in the mail for "sales literature," Mr. Packard told his secretary, "We don't have any sales literature. Tell him, oh, tell him to ask the man who owns one," and it stuck.

Enter, Henry B. Joy, an unexpected visitor at the New York Auto Show in 1901. He was a businessman from Detroit with a vast railroad fortune. He came to New York in search of a motor for his boat. He was impressed with Packard's quality as he witnessed two of them start up quickly to follow a screaming and puffing fire engine. He bought two model C Packards plus 100 shares of stock. In January, 1902, he bought 150 more shares of Packard stock. In October, 1903 the company moved to Detroit where its name was changed to "The Packard Motor Car Company," and it made a stock offering of $250,000

Henry Joy with some friends subscribed to $150,000 of this offering thus shifting control to Henry B. Joy and Associates. Joy became president and held this position until 1916 when he resigned over a dispute about a merger with Nash.

Survival is always a haunting question. The race is not always to the swiftest. Packard hit the ground running and was almost an instant success. It became one of the best known luxury cars in the world with early sales in Britain, Cuba, France, South America, and Spain. In some years Packard outsold Cadillac. With Packard's phenomenal success, prestige and image, why should its stable yield to lesser steeds? Some say it was a casualty of World War II, never able to regain its glory days. No organization is better than the people involved and their ability or genius (or lack of it) to cope with the vagaries of history and the limits of the human condition. Steve Smith in his history of Packard concludes it was not the quality factor that sank Packard but business practices and bad decisions.

A personal note: In 1940 I parted with a 1933 Chevrolet and $350 for a used grey 1937 Packard 120 straight eight, four-door sedan. It served me well until 3:30 one morning in northern Virginia in 1943. A telephone pole fashioned a "V" out of the front bumper and did a number on the steering and suspension underneath. It was never the same after that.

Henry B. Joy

Special Collection, University of Michigan Library

PACKARD FIRSTS

Steering wheel 1901

H gear-shift configuration 1901

Automatic spark advance 1899

Under a mile-a-minute 1903

Handbrake on left side of driver 1915

American twelve cylinder 1915

Aluminum pistons 1915

Hypoid differential 1925

Hydraulic shock absorbers 1926

Back up lights 1927

Pressure cooling system 1933

Power hydraulic brakes 1936

Air conditioning 1939

Sealed beam headlights 1939

Automatic windows 1940

All steel station wagon 1948

Special Collections , University Of Michigan Library
Gail Hoag

Gail Hoag, field secretary of theLincoln Highway Association and the LHA Packard, 1926. A symbol of the association, the powerful car represented grace, pride, speed, and things done the right way.

INDIANA
And The Lincoln Highway

Miles: 154.3 Towns: 20

LINCOLN HIGHWAY HISTORY IN INDIANA is quite straightforward—not so torturous as in Ohio, Nebraska and other states. Indiana, however, is one of the most significant Lincoln Highway states, particularly through the "eyes" of an automobile. After all, the car is what it's all about. Right?

Over 350 marques were produced in Indiana and

Special Collection, University of Michigan Library

The LHA 1915 Packard on the Ohio—Indiana state line.

of these, seventy-one were manufactured in Lincoln Highway towns like Fort Wayne (13), Ligonier (1), Goshen (3), Elkhart (30), Mishawaka (6), South Bend (14), and La Porte (4). Some names like Cord, Duesenberg, Studebaker and Packard have been indelibly etched in automotive history, even though none have survived.

Indiana is in questionable running for the distinction of producing America's first automobile in 1894. The main competition was the Duryea in Massachusetts in 1893, but Elwood Haynes claimed that the 1893 Duryea was only a motorized buggy not a bona fide automobile. Haynes-Apperson also began in 1893, when Haynes purchased a one-horse Sintz engine at the Chicago World's Fair. His first experimentation with the engine was in his kitchen, mounted on sawhorses, in Kokomo,

Indiana. His Pioneer car saw the first light of day on Pumpkinvine Pike in Kokomo on July 4, 1894. The roads of Indiana and beyond saw many Elwood Haynes' cars for the next thirty-one years, until production ceased in 1925. Haynes and Apperson split in 1901. The oldest Apperson is an 1895 Pioneer II at the City of Firsts Automotive Heritage Museum in Kokomo, Indiana.

The most noteworthy auto birthed on the Lincoln Highway has to be the Studebaker. Its history is intriguing and truly romantic. The saga begins in a small blacksmith shop in 1852, which ultimately became the world's largest wagon manufacturer after the Civil War. Studebaker autos began as electric vehicles in 1902 and continued as such until 1912.

Gasoline powered Studebakers arrived in 1903 and continued to be produced until 1963, when the Studebak-

Elwood Haynes Museum, Kokomo

Haynes Pioneer, 1894

Elwood Haynes sits in his 1894 Pioneer

er company could no longer compete as an independent with the Big Three. The South Bend Company competed with head high and with no apologies. From 1901 to 1910 Studebaker ranked fourth of all American producers. In 1911 it was second only to Ford, and from 1912 to 1914 was third behind Ford and Willys-Overland. It survived the Depression and World War II and in 1962 gave birth to the Avanti, a personal luxury car.

It's important for Lincoln Highway fans to know that it was through the efforts of Clement Studebaker, Jr. that South Bend became host to the Lincoln Highway. An item in the South Bend *News-Time*s on Sunday, September 14, 1913 reads:

> Through the efforts of Clement Studebaker Jr., of the Studebaker Corporation, South Bend was placed on the Lincoln Highway, the route of which was announced today. By persistent and financial donation to the building fund of the national road he was able to show the committee in charge the desirability of having South Bend on the route. Although no definite local route has yet been announced it is believed the highway will pass through the heart of the city, entering it on the east possibly along Jefferson boulevard, from Mishawaka and thence west out along the Michigan avenue road to Laporte. New Carlisle may also be touched by the highway. It is estimated the highway will cost over $10,000,000, which will be provided by popular subscription. Already $5,000,000 is pledged. The road which is to be constructed of concrete whenever practicable, will reach from New York to 'Frisco and will be open to lawful traffic of all descriptions. No toll charges are to be paid.

What do Indiana, the Lincoln Highway and the Indianapolis Motor Speedway have in common? Answer: Carl G. Fisher.

> Fifteen-year-old Indianapolis resident Jane Watts was walking along Meridian Street one fall afternoon in 1908 when she noticed something strange. All traffic on the street had stopped and people were craning their necks upward. Following their lead, Watt stopped, looked up, and was stunned to see a giant hot air balloon floating by with, instead of the usual wicker basket, a Stoddard-Dayton automobile. Sitting in the car she saw, for the first time, the man she would marry—Carl G. Fisher.

Thus begins Indiana Historical Society's brief history of Carl G. Fisher. This stunt to promote the Stoddard-Dayton auto was typical of the lavish imagination of the "Hoosier Barnum."

What Jane Watts and the public did not know, was that Fisher had removed the engine to make the car easier to lift. His brother Rolly had driven a similar car to the landing site for Fisher to use for his triumphant return to Indianapolis. Jane Fisher later said, "It always puzzled Carl that no one had been suspicious enough to follow his flight and that the public, press, and police had been so easily hoaxed."

Fisher was born January 12, 1874, in Greensburg, Indiana. At 12 he quit school because of severe eyesight problems, got a job in a grocery store and declared to his single-parent mother: "From now on, I'm supporting this family." At 17 Fisher and his two brothers opened a bicycle shop in Indianapolis, which was proclaimed as the "Finest bicycle shop in Indiana." As bicycle popularity yielded to the automobile around the turn of the century, the Fisher bicycle shop was converted to auto repair and sales.

In 1904 Fisher initiated the Prest-O-Lite Company to manufacture automotive headlights by a patented system of compressed gas. The successful venture finally boasted factories in Indianapolis, Cleveland, Omaha, New York, Boston and Chicago, and in 1911 was sold to Union Carbide for $9,000,000. In 1909 Fisher, along with friends Allison, Newby, and Wheeler, created the Indianapolis Motor Speedway Company with $250,000, creating a

University of Michigan

Carl G. Fisher

two-and-a-half-mile oval that has become synonymous with auto racing. The inspiration for this came during a visit to Europe in 1905. Fisher was stunned by the superiority of European cars over American makes. "They can go uphill faster than American cars can go down," he said. He conceived of a proving ground where cars could be tested. On August 19, 1909, the first auto races—300-miles were run at the speedway with less than happy results. Six people were killed and Fisher stopped the race after 235 miles. The first track was paved with crushed stone, the second with 3,200,000 ten-pound bricks – thus the "Brickyard" was born. On Memorial Day 1911 the first 500-mile race was run. This gave birth to a long line of races which continue to the present. That first "500" was won by an Indianapolis-made Marmon Wasp, with an average speed of 74.59 mph.

The next project to emerge from Fisher's energetic imagination was the coast-to-coast highway. Fisher was hooked on automobiles and it took no crystal ball to see that a basic need was good roads upon which to travel. He saw that whatever was good for cars was good for the wallets of those who manufactured them, sold them, or made parts and provided accessories and service for them. Good roads were as basic as one could get. Fisher became totally committed to what became the Lincoln Highway, a memorial to Abraham Lincoln.

At a September 1, 1912, dinner party for auto manufacturers at the Deutsches Haus in Indianapolis, Fisher waxed eloquent, "A road across the United States! Let's build it before we're too old to enjoy it." Just thirty minutes after his talk he received a $300,000 pledge from Frank A. Seiberling of the Goodyear Tire and Ruber Company and shortly thereafter a $150,000 pledge from Henry Joy, president of the Packard Motor Car Company. On July 1, 1913 the Lincoln Highway Association was created with Joy as president and Fisher as vice president. Fisher had a good sense of timing and the Panama-Pacific International Exposition in San Francisco, commemorating the completion of the Panama Canal, did not go unnoticed by him.

As the Lincoln Highway gained its own momentum, Fisher typically turned to other mountains to climb – like Miami Beach in Florida. Miami Beach was not a mountain but a mangrove swamp, where the visionary Fisher saw a potential vacation spot for midwestern automobile executives to escape the frigid winter weather. Fisher built Miami Beach. An adjunct to the Miami project was the Dixie Highway, not like the east/west Lincoln Highway linking coast to coast, but a north/south highway linking Chicago to—where else?—Miami Beach. On December 4, 1914 Fisher wrote to Indiana Governor Ralston with this Dixie Highway proposition. On April 3, 1915 the governors of Illinois, Ohio, Kentucky, Tennessee, and Georgia, at Governor Ralston's invitation, met in Chattanooga to discuss the proposition. In September 1916 Fisher and Ralston attended a celebration in Martinsville, Indiana, opening the roadway from Indianapolis to Miami.

Having conquered a Miami swamp, Fisher turned to a new development, even larger than Miami Beach, at Montauk, at Long Island's eastern tip, but this time history did not smile on him. His multi-million dollar

Henry Ostermann Memorial, 1921-1923

This memorial accomplished two things: It memorialized Henry C. Ostermann, vice president and field secretary of the Lincoln Highway Association, who was killed on the Lincoln Highway in 1920. It also identifies the location of the "Ideal Section" of the Lincoln. This 1-1/3-mile section of Indiana road between Schererville and Dyer received world-wide attention as promoters publicized the project in search of the best materials and engineering. This sparked a debate between concrete and brick, with concrete emerging the victor. Right-of-way was 100 feet with paving forty feet wide and ten inches thick. It was "ideal" for its time, but was judged a few years hence as somewhat underwhelming. Its engineering assumed average auto speed of 35 mph and trucks at 10 mph. However, it was a great beginning and we are all beneficiaries of that vision and commitment. The memorial is on the south side of U.S. 30 (Lincoln Highway), 1.3 miles west of the traffic light at U.S. Highway 41.

Special Collections, University of Michigan Library

Ideal Section, *Early Four-Lane "Ideal" Section of Lincoln Highway at Dyer, Indiana*

investment came to haunt him with the devastating hurricane of 1926 and the Great Depression in 1929. He was forced to sell his Miami property to satisfy Montauk bondholders. Walter Myers, Fisher's long time attorney recalls his last visit with Fisher when, on a visit to Miami, he spotted Carl standing with one foot on a park bench. He asked him how he was doing and received a non-Fisher like reply:

> I can tell you in a few words. The bottom dropped out of the sea. New York and Long Island took everything I had. I'm a beggar—dead broke, no family to fall back on. Yes, the bottom dropped out of the sea and I went with it. You know, I promoted Miami Beach here. The grateful people got up a purse, five hundred dollars a month for me. That's what I live on. I used to make dreams come true. Can't do it any more. I'm only a beggar now. The end can't be far away.

Fisher died from a gastric hemorrhage on July 15, 1939, in Miami Beach, Florida.

Indianapolis Motor Speedway, Trackside Photo Shop.

First "500"— Indianapolis, 1911

Start of the first Indianapolis 500 race, May 30, 1911. Carl Fisher's personal Stoddard-Dayton roadster is on the pole, with partner James Allison as passenger. Next to the pace car is the #1 Case driven by Lewis Strang. #2 is De Palma's Simplex. #3 Harry Endicott's Inter-State. #4 is Johnny Aitken's National, and behind it is the #9 Case driven by Will Jones.

Brick Remnent, *south of Ligonier, Indiana. Note the unusual placement of the brick to form the curb.*

Columbia City, Indiana

108 Jefferson Street in front of the Whitley County Historical Museum, Columbia City, Indiana.

History Center

This view of the post in Fort Wayne is in the History Center where it was on loan from the Lincoln Center Museum in 2003

Columbia City, ~~Warsaw,~~ Indiana

9602 on old Lincoln Highway, eight miles east of Route 30, Warsaw, Indiana.

Warsaw, Indiana

602 Lake Street in Funk Park,

Hamlet, Indiana

Beneath this elevator in Hamlet, Indiana lies a post. A Lincoln Highway friend noticed a prostrate concrete post, in the way for the construction of a new elevator. He inquired about the posts future and was told he could have it if he came back later that afternoon. Upon his return he was informed that the post had been thrown into the foundation fill for the elevator.

Donaldson, Indiana

One mile east of the Lincoln Highway intersection with Union Road on south side. house number, 19473. It is partially hidden by a bush at the driveway entrance,

Hamlet, Indiana

208 Stack Street on the west side of the street,

Donaldson, Indiana

Dr. Miller's son lifted an upturned canoe along the side of the house to reveal this partial post he rescued from a nearby trash heap. They plan to "plant" it next to the post by the hedge.

Hamlet, Indiana

508 West Indiana Avenue, in a grove of trees close to the house.

FORT WAYNE, *Indiana*

IN HIS OWN WORDS:

Wednesday, August 18, 1915

Arrived in Fort Wayne, Indiana 81 miles. Rough and muddy roads

FORT WAYNE, PORTAL TO INDIANA became important to the Lincoln Highway as the gateway to the auto industry and thus to the Lincoln Highway. Fort Wayne was home to the manufacture of eight different autos plus six more that died in childbirth. Of the eight bona fide marques, seven did not add up to even a blip on the auto screen: Black (1906), Columbia and Eclipse (1902), Economy (1909-1911), Roach & Albanus (1899-1900), Slattery Electric (1889), and TJK Special (1909). But Fort Wayne hit it right with the eighth—Duesenberg (1920-1937) whose name even now generates automotive goose pimples.

Duesenberg history has its roots in a 1903 bicycle company bankruptcy in Iowa. Fred Duesenberg then partnered with brother Augie to build race cars. This led from Iowa to St. Paul, Minnesota, to Elizabeth, and Newark, New Jersey, to Indianapolis, Indiana where one day in 1926 E. L. Cord walked into Fred's office, acquired the company and instructed Fred to build a super-car to stand toe-to-toe with any of the world's most magnificent automobiles.

Duesenberg became more than a status symbol; it was status. Always upscale in quality, performance and price made it an easy victim of the unforeseen stock market crash in 1929. Fred Duesenberg died in an auto accident in 1932, and Duesenberg died along with Cord and Auburn in 1937. The *Catalogue of American Cars* summarizes Duesenberg, "If but one automobile ever built in America had to be singled out as the most glorious achievement in the country's automotive history, that car would have to be the Duesenberg. It transcended the ordinary in full measure, created legends in its wake which will live forever, and became a literal metaphor —'It's a Duesy.'"

As with many other American towns, Fort Wayne's history was built upon the history of American Indians. For Fort Wayne it was the Miami Indians, and specifically Chief Little Turtle, born in 1747 not far from the site of Fort Wayne. Little Turtle, chiefly noted for his brilliant military career, was a half-Miami and half-Mohican Indian but became a leader among the Miami Indians. Though far ahead of his time in military affairs, he concluded that war was not the way to settle difficulties. Since Indians believed that land was held in common, he opposed the ceding of land to the white men and was the last to sign the Treaty of Greenville in 1795. He said, "I was the last to agree to make this treaty; I shall be the last to break it." He kept his word, even though the Americans did not.

Chief Little Turtle was granted many favors by the U. S. government, and both George Washington and Thomas Jefferson received him at the Capitol. Although Little Turtle recommended to his people that they avoid the ways of civilization, he, himself, ironically died at age 65 in 1812 of a disease of civilization—gout.

Fort Wayne is also host to the grave of John Chapman, better known as "Johnny Appleseed." He was born September 26, 1774, in Massachusets but traveled the Midwest as a missionary and orchardist. He is remembered by all as the man who planted apple trees all over the Midwest. He died March 18, 1845, at 71, and is buried in Archer's Graveyard, Johnny Appleseed Park, in Fort Wayne, Indiana.

History Center

The post is in the History Center where it was on loan from the Lincoln Center Museum, 6-13-03.

Harrison Street Bridge

Harrison Street Bridge over the St. Mary's River in Fort Wayne.

Old Iron Bridge

Old Lincoln Highway Bridge just north of Fort Wayne.

Wells Street Bridge

The original Wells Street Bridge over the St. Mary's River in Fort Wayne.

LA PORTE, *Indiana*

IN HIS OWN WORDS:

Thursday, August 19, 1915

From Ft Wayne to LaPorte, Ind. 105 miles.
Good road— some mud.

THE FIRST DOWN PAYMENT on the Lincoln Highway was not a donation from some auto manufacturer. It was from the indigenous people who lived here centuries before. There is a recurring theme in the towns of midamerica hosting the Lincoln Highway that American Indians played a pivotal role, often actually providing the first trail sites. Unfortunately, the Indians' final contribution was involuntary. The scenario is always the same, differing only in details and scope. The land belonged to everyone as does the air; white men made treaties for use of the land; white men broke treaties; white men took ownership; Indians were displaced.

The story in northwest Indiana is stark, (ironically one of the first counties was called "Stark") and devoid of any honor. Before 1830 all of La Porte and Stark counties was part of the Pottawatomie nation – a peaceful people with trails and traces that ran through the forests, around the lakes and along rivers and creeks. In 1838 President Martin Van Buren, following the tradition of his predecessor, Andrew Jackson, removed the Indians from northern Indiana to Osage County, Kansas. This was a long thousand-mile journey by foot, with so many Indians dying en route that it was called the "March of Death."

On such a somber note, let's turn to things more pleasant, like La Porte, the town. It was founded July 1832 by A. P. Andrew, James Andrew, Dr. Hiram Todd, John Walker, and Gen. Walter Wilson. They were followed later by immigrants from Sweden, Germany, Poland, and Italy. The name "La Porte", meaning "The Door" was given to the area because of an opening in the trees (probably maples). La Porte is nestled in an area surrounded by magnificent lakes carved out by glaciers during the last Ice Age, 10,000 to 15,000 years ago. As early as the 1830s, Hoosiers were improving their roads with glacier deposited gravel. Today, nearly $40 million worth of gravel and sand are extracted annually in northern Indiana.

Through the eyes of the Lincoln Highway, La Porte is an auto town, stepmother of the fantastic Munson. John W. Munson of Chicago was the genius behind

LaPorte County Historical Society, Inc.

The Munson Company, La Porte, Indiana

70

1898 Munson Omnibus

Shown in the photograph from left to right: Edwin Beeman, John H. Munson, the inventor, Miss Emma Porter (back), Helen Andrew Patch, Mrs. Jennie Stahl Powell, Mrs. Elsie Haggard Ridgway, Mrs. Pearl Ridgway Grandstaff (below), Mrs. Freda Mayne Chaney (above), Mrs. Pearl Hewlett Hall (below.)

the Munson in 1896. About 1899 the company moved to La Porte, built four cars and produced a beautiful brochure to help promote the car to Chicago and eastern capitalists. The seven-passenger Omnibus made two trips to New York and was the first automobile to drive along the famous Riverside Drive. Details are murky leading to the Munson demise, which was unfortunate because the car enjoyed genuine support and loyalty from the town and for good reason: Munson was born 100 years too soon. Believe it or not, the Munson was America's first hybrid, combining the best virtues of both gasoline and electricity. "The electric machine operates automatically, and is either a generator or a motor, according to the speed of the engine." The other before-it's-time innovation was the self-starter. A Munson Company advertisement explains how it works:

The machinery is started by moving the controller lever, located in front of the driver's seat. It sends a current from the storage battery through the electric machine and starts the vehicle. The gasoline engine immediately begins to explode and reverses the current through the motor, that is to say, the motor then becomes a generator, and sends a charging current into the storage battery.

NO SECONDARY BATTERY, CELL, OR DYNAMO IS USED FOR SPARKING PURPOSES.

The gasoline engine delivered 12 hp with four speeds. The advertising claimed that,

10 gallons of gasoline with constantly charged storage batteries, will furnish power to propel the vehicle 100 miles or more over the ordinary, well-traveled, country roads, at the rate of five to fifteen miles per hour, according to the conditions and gradients of the roads.

Buggies of Any Description

The Munson Company of LaPorte, Indiana; speeds, 2, 4, 8, 16 miles per hour; Ample luggage room.

PRE-WWII U.S. AUTO MANUFACTURING

"The history of the American automobile is a can of worms that would make any self-respecting bilateral invertebrate blanche"

THUS SAID AUTO HISTORIAN AND author, Beverly Rae Kimes. If auto marque tabulation is a wormy enterprise to such an auto expert as Beverly Rae Kimes, it hardly behooves me to indulge in the postulation of such a slippery subject. However, it does seem fair to say that in the years prior to WW II, 5,000 different automobiles were built. To quote Kimes again;

> The American automobile did not arise full-grown onto an assembly line. Its borning years and its puberty were experienced in hundreds of villages and towns from coast to coast, as small machine shops, bicycle builders and carriage makers – even doctors, jewelers and florists – used a few tools and varying talent to put themselves automotively on the road. Some of these early pioneers built their cars with the glorious hope of fullfledged manufacture, others because they thought they could do better than what was available to them on the market, which often they could, as witnessed by some of the admirable one-off efforts which remain in existence to this day. Other attempts were less successful, of course, and some were downright bizarre, but all of these pioneer vehicles deserve to be woven into the fabric of this nation's automotive heritage because they were undeniably a part of it.

It's important to think about this because the cause and effect are intertwined between the Lincoln Highway and the automobile. It takes little reflection to note the dramatic evolution of the auto—from 5,000 marques (400 manufacturers) to less than 10 today, and those from only three U.S. manufacturers: Ford, General Motors, and Daimler/Chrysler. To stir the worms even more, the very name, "Daimler/Chrysler", identifies the crossbreeding within automotive stables. Specifics today will be outdated tomorrow – evidence we live in an increasingly global world. Forty manufacturers at the first auto show at New York in November 1900, displayed more than 300 vehicles. The five states with the highest auto production were: New York, 845; Michigan, 592; Ohio, 545; Illinois, 433; Massachusetts, 393. Between 1913, the birth of the Lincoln Highway, and 1997, deaths per 10,000 vehicles were reduced from 24 to 2. The mileage death rate in 1927 was 18.2 deaths per 100,000,000 vehicle miles, while in 1997 the death rate was 1.74 per 100,000,000 miles, the lowest rate on record. This testifies to enormously safer cars as well as safer roads and, road rage and cell phones notwithstanding, even safer drivers.

ILLINOIS
And The Lincoln Highway

165/179 miles (1924/2001) 18 towns (1915)

ILLINOIS IS FAMOUS FOR MORE THAN hosting the second largest U.S. city—Chicago. Rolling prairies, urban cityscapes and historic riverfront towns are all encountered on the Lincoln Highway in Illinois. In fact, my father drove over the very first Lincoln Highway "seedling mile" in 1915. In those early days of the auto and roads, the thing most often encountered beyond the city limits was mud. The Lincoln Highway Association volunteered to lead the way in providing good roads between towns. They began by paving strips of show-and-tell pavement and called them "seedling miles," with the expectation that this good example would catch the imagination of road supervisors, farmers, politicians and any other mover or shaker who could cultivate those "seedlings" to grow everywhere, connecting town with town and state, with state in triumph over mud. Please note: Illinois was the first state to achieve this goal. It began with the first seedling mile in October 1914, on the Lincoln Highway just west of Malta, Illinois. In 1915 an additional seedling mile was paved in Morrison.

But before popping any vest buttons, the Lincoln Highway was not the inventor of short strips of good road. The honor goes to Gen. Roy Stone, the country's first federal road agent under the new Office of Road Inquiry, who built the first one-quarter mile "object lesson" road in 1896 in Atlanta, Georgia. Within three years he planted 21 object-lesson roads in nine states.

Paved roads in Illinois were no historical accident.

Franklin Grove, Illinois

Lynn Asp and the Franklin Grove post outside the National Headquarters of the Lincoln Highway Association. Lynn is manager of the National Headquarters office.

It was a matter of "putting your money where your mouth is," though documenting the achievement may not be entirely possible. Nevertheless Illinois was a significant player on the stage of auto production and spawned 433 different marques with twenty produced in four Lincoln Highway towns. Of the total, 238 autos came from Chicago, but Chicago was not a Lincoln Highway town. The Lincoln skirted Chicago to the south and west in order to avoid city congestion. Now this was 1913—city congestion? For the record, the July 9, 1914, issue of *Automobile* reports that auto registrations in the state of Illinois alone were 110,000 with 1,203,770 registrations in thirty-three other states! "The more things change, the more they stay the same."

The automobile producing towns along the Lincoln Highway in Illinois were: Aurora *eleven*, Batavia *one*, Sterling *six*, and Chicago Heights, the most important (granted a biased opinion), *two* cars and *one* truck. The truck, of course, was the Little Giant, one of which was driven by my father across the Lincoln Highway to San Fransisco in 1915.

The history of Illinois goes back further than any of us knew to go. It goes back to 10,000 B.C., when the Paleo Indians hunted mastodons and crossed streams swollen by melting glaciers. By 800 to 600 B.C., the Archaic Indians occupied the area, followed by the Woodland Culture Indians, 600 B.C. to 800 A.D. A people who added decorated pottery and tools for agriculture to their

making of hunting tools. Between 800 to 1,300 A.D. the Mississippi Indians started growing beans and, corn and squash, and so improved agriculture as to create an environment attractive to the French in 1673, when Jacque Marquette and Louis Joliet decended the Mississippi River.

It's a bit of a stretch to connect the Lincoln Highway with Indian life 8000 years ago. Nevertheless, under the hypnotic effect of driving the Lincoln Highway across those rolling prairies of northern Illinois and the spell of golden waving grain, it is somehow comforting knowing that it has been this way a long time and we are treading where not a few have gone before.

Franklin Grove

Miriam and the author stand in front of the Franklin Grove Lincoln Highway headquarters

New Lenox, Illinois

In the lawn of the Lincoln Way North High School, north side of the Lincoln Highway. The post was relocated in May, 1996

Ashton, Illinois

State Road 38, 3.8 miles west of Ashton on north side of road.

Ashton, Illinois

State Road 38, 3.6 miles west of Ashton on south side of road.

Ashton, Illinois

State Road 38, 2.4 miles west of Ashton on south side of road

Franklin Grove, Illinois

Lyn Asp and Miriam at the Franklin Grove Post.

Ashton, Illinois

State Road 8 on curve 2.3 miles west of Ashton on north side of road.

CHICAGO HEIGHTS, *Illinois*

IN HIS OWN WORDS:

Friday, August 20, 1915

Rain. From LaPorte, Ind to Chicago Heights, Ill., Chicago Pneumatic Tool Company plants, 55 miles. Took train into city.

Saturday, August 21, 1915

Rain. Spent the day in Chicago, Ill.

Sunday, August 22, 1915

Fair. Spent the day in Chicago, Ill.

IT IS THE STORY OF AN "American Dream Come True". The Little Giant truck was manufactured by the Chicago Pneumatic Tool Company of Chicago Heights, Illinois, the progeny of John W. Duntley a foundry foreman in Milwaukee, Wisconsin. Duntley had vision. The year was 1889; the United States was in its devastating recession of the late nineteenth century. Work at the foundry where Duntley was employed was reduced to only one day a week, he went on the road and launched an aggressive sales program. His salesmanship kept the foundry in full operation until the end of the depression nine years later. His success convinced him of the potential for producing industrial products, so in 1894 he developed his own sales agency which he called "Chicago Pneumatic Tool Company," manufacturing chipping, riveting and scaling hammers. Due to a series of fortuitous events, manufacturing began in Saint Louis, Missouri, where it experienced very rapid growth. A peek behind the company curtain reveals some interesting trivia:

The office will be open from 7:00 A.M. to 8:00 P.M. daily except the Sabbath, on which day it will be closed. Men employees will be given an evening each week for courting purposes or two evenings a week if they go regularly to church. Bring a bucket of water and a skuttle of coal for the day's business.

From 1901 to 1913, Chicago Pneumatic Tool Company had plants in Utica, Detroit, Cleveland, Chicago, and Philadelphia, in Montreal, Canada; and overseas in London, Berlin, and Scotland.

In 1911 the Franklin, Pennsylvania, plant ventured beyond the world of industrial tools to produce the Little Giant truck, a gasoline, 2 cylinder, chain driven light

Chicago Pneumatic: The first Hundred years

An early Little Giant taxi cab, Washington, D.C.

truck. Within two years production reached the respectable figure of 100 trucks per month. Enter my father. He was bitten by the "Westward Ho" bug and determined to drive the Lincoln Highway to San Francisco, but on March 18, 1915, he went truck hunting in Philadelphia and kicked the tires of a Willys Overland truck. On April Fool's day, 1915, he took a demonstration drive in a Little Giant truck in Pottsville, Pennsylvania, and on April 5, 1915, he went to Philadelphia in a late spring snowstorm and purchased his Little Giant for $1,240, including a ten percent discount. On Friday June 4, 1915,

February, 1916, Chicago Pneumatic.

A Little Giant with special tank body in the service of Standard Oil Co. On account of its reserve power the Little Giant makes a fine showing on rough, muddy roads, and in hilly country where the going is difficult. The truck is covering a large expanse of territory in the vicinity of Polo, Illinois, and is giving excellent account of itself.

he pointed the Little Giant westward on the Lincoln Highway and headed for the Panama-Pacific International Exposition in San Francisco.

The Little Giant, hardly a blip in the annals of truck history, enjoyed a place in the stables of some important owners like the Coca Cola Company and the Tiffin Consolidated Telephone Company. The crowning achievement for the Little Giant was to be a member of that select quintet assembled to produce the "Three-mile Picture Show" in 1915, a 16,000-foot, three-hour movie filmed by the Lincoln Highway Association, en route from New York to San Francisco. The Little Giant was provided by its parents, the Chicago Pneumatic Tool Company, to help with logistics. Its bedfellows were a Stutz, Studebaker, Packard, and Oakland.

Among other firsts in 1915 was the inception of the taxicab industry. Eventually, twenty-eight Little Giant trucks were committed to the wild notion that people would be willing to pay money for an automobile ride. The Arlington-Barcroft Co. of Washington, D.C. put the Chicago Pneumtic Tool Co. vehicles into service as the "Yellow Jitney Line." These "buses" had padded seats along the sides and rear. The charge was a "Jitney" (a nickel).

"The rugged Chicago Pneumatic Commercial Car" in 1913 became the second largest department of the company. By 1914 there were 25,000 Little Giants in service and in the same year the Truck Division moved from Franklin, Pennsylvania, to Chicago Heights, Illinois, where manufacturing continued until 1918. In that year the division was sold and the manufacture of trucks discontinued, because trucks were thought to be out of sync with the company's other products.

The company finally moved its offices to Rock Hill, South Carolina and shortened the name to "Chicago Pneumatic." During WWII the company became famous for its riveting equipment, once shown on the cover of the March 29, 1943 issue, of *The Saturday Evening Post Magazine* with "Rosie The Riveter."

LINCOLN HIGHWAY FROM GOTHAM TO FRISCO
ROADWAY WILL TOUCH CHICAGO HEIGHTS
SAUK TRAIL PART OF NATIONAL HIGHWAY

were the headlines on page one of the *Chicago Heights Signal*, September 13, 1913. These were typical headlines in Chicago Heights papers in late 1913—the founding year for the Lincoln Highway. As in many potential Lincoln Highway cities, Chicago Heights was a participant in the dialogue concerning the route of the highway. The final decision was made at a special meeting at the Elks Club on Friday, November 27, 1913, with challenges coming from Crete and Frankfort Station. The special guest was A. R. Pardington, vice president and secretary of the Lincoln Highway Association, who helped bring to conclusion the choice of route in this vicinity. Two primary factors determining the choice of the route were the list of state aid towns, and the judgment of the Chicago Heights Automobile Association. Following extensive discussion and careful inspection of state aid maps, Chicago Heights became the central focus of those present. After adjournment, Pardington was entertained at a dinner at Hanson's by President Wm. Waddington of the local auto association.

GENEVA, *Illinois*

IN HIS OWN WORDS:

Monday, August 23

Fair. Left Chicago Hts. at 1 P.M. Arrived in Geneva, Ill. 57 miles. Roads very rough

All Aboard! Chicago & Northwestern Railroad Depot in Geneva at the turn of the century

THE YEAR 1830 WAS A VERY GOOD ONE for Geneva. This is the year white settlers first entered the stage of this northern Illinois community. The first, Daniel Shaw Haight, of Dutch origin, built his cabin near a spring along Fox River in 1833. The "settlement" was called "Big Spring." Two years later Haight sold his claim to James and Charity Herrington, to whom Margaret was born, the first child born in Geneva. Early names for the community, besides Big Spring, were Herrington's Ford, La Fox, and Campbell Ford. In 1836 the community was selected as the Kane County seat, at which time the name was changed to Geneva, most likely at the suggestion of influential Dr. Charles Volney Dyer, a noted abolitionist who had recently come from upstate New York, where Geneva was a common name.

James Herrington died at age 41 (an untimely death) but he was more than the second signature on the first Big Spring claim. During his short tenure he plotted the town, helped establish the county seat, was elected county sheriff, and opened the first general store, tavern and post office. It's interesting to note that in spite of his contributions to community infrastructure, the name "Herrington" does not appear on the sign at the city limits. It is Geneva.

In 1836 the "Boston Colony" arrived from Massachusetts and were instrumental in organizing the Unitarian Church, the oldest church in Geneva, built in 1843. This church was shortly followed by Methodists, Congregationalists, Swedish Lutherans and Disciples Of Christ. History has shown that churches were intrinsic components of the early towns along the Lincoln Highway.

By 1840 Geneva, just forty miles west of Chicago, was on its way to becoming a really significant community. It could now boast a courthouse and jail, a post office, a classroom and teacher, a bridge, a sawmill, at least three general stores, a doctor, a furniture and coffin maker, at least two blacksmiths, two hotels, and a tavern. In 1858 it was incorporated as a village, and in 1887 as a city. One of the most important developments, as it was for most Lincoln Highway towns, was the coming of the Western Pacific Railroad in 1853. This established a vital relationship between Geneva and Chicago and greatly helped create the symbiotic relationship between the Lincoln Highway and the railroads.

Geneva History Center

Ready To Go! 1915
The Geneva Commercial Club leaving for a tour to Delavan, Wisconsin

When my father drove through town on August 23, 1915, he was probably inconvenienced by road construction because this was the year for paving the Lincoln Highway (State Street) and Third Street. In fact, his comment was, "Roads very rough." In the same year Henry Fargo built his "Fargo Block" on west State Street and my father certainly passed by it as he putt-putted west on that early edition of the Lincoln Highway. In the latter part of the twentieth century, the Fargo Block was occupied by the Henry Store, City Barber Shop, Anderson Jewelers, and Filbert Drug Company.

Geneva, like many other midwest Lincoln Highway towns, profited from the great variety of ethnic settlers. Two widely diverse groups became the main ingredients of Geneva's "fruit salad." The largest group came from Sweden to help construct the railroad. Many were so impressed they returned to settle down as permanent residents. The second largest group came from Italy, to operate their own businesses or work on the railroad. Both the Swedes and Italians, along with smaller groups of German, Scottish, and Chinese heritage, made their own unique contributions to the rich culture of Geneva.

Geneva History Center

All-Weather Street

The Lincoln Highway (State Street) looking west from Second Street, is readied for brick paving in 1916. Beneath the concrete surface of today lie the original paving bricks. Note interurban tracks

Modern Geneva is a charming town of more than 19,000, with a wonderful blend of agriculture, shops, recreation, and businesses, committed to a wholesome respect for history, responsibility for the present and vision for the future.

FULTON, *Illinois*

IN HIS OWN WORDS:

Tuesday, August 24, 1915

Fair. Traveled from Geneva, Illinois to Fulton, Ill. 109 miles. Roads very rough

FULTON IS A MISSISSIPPI RIVER TOWN. Its first settler was John Baker, who in 1835 chose the bottomlands which were good for hunting, and where the Mississippi River was narrow enough for easy crossing. It was Baker who had the vision of a city here on the Mississippi. Twenty years later, in 1855, this became a reality when Fulton was incorporated. The state legislature approved a special charter for the city of Fulton in 1859. Some of the buildings of that era are still occupied.

Early community development most often depended on two things: bridges and railroads. For many Lincoln Highway towns the two depended on each other as in Columbia/Wrightsville, Pennsylvania, and again here in Fulton. The ferry dates from the very beginning of Fulton, powered first by man, then horse, then steam. In 1891 the Fulton-Lyons high bridge vastly improved transportation between Illinois and Iowa.

Kevin Heun

Fulton-Lyons High Bridge

1891 bridge over the Mississippi River at Fulton, Illinois. Prior to this bridge, travel was by ferry.

Thirty-six years prior to the coming of the bridge, the railroad arrived and was greeted with great excitement and optimism. The mammoth Dement House Hotel opened in 1855, the same year that the Galena and Chicago Union Railroad arrived in Fulton. In anticipation of this event and its hoped-for stimulus for growth and prosperity, Charles Dement spared no expense in the construction of this five-story hotel with 23-inch-thick native limestone walls and lavish furnishings. The opening was a gala event with a band playing in the roof garden and everything a "blaze of gaslight and glory". Unfortunately, the bridge, to complement the railroad, did not arrive until 1891. The railroad alone could not provide sufficient economic stimulus and in 1861 the Dement Hotel was forced into bankruptcy after only six years of operation. This grand building, created with such great hopes and dreams, became a military school for Civil War soldiers, headquarters for disabled veterans, Northern Illinois College, Northern Illinois Military Academy, and for several years from 1916 to 1923, home to the Lincoln Highway Tire Company. The structure was razed in 1935.

Fulton is also known for its celebrities. Many of President Ronald Reagan's relatives are buried here. Michael O'Regan, the president's great-grandfather moved from Tipperary County, Ireland, to Carroll County, Illinois in 1858. His son John, Ronald Reagan's grandfather, moved to Fulton to work in a grain elevator. Ronald Reagan's father worked at the I. W. Broadway Dry Goods Store, and was married to Nellie Wilson at the Fulton Catholic Church on November 8, 1904. The Reagans moved to Tampico in 1905, where Ronald was born on February 6, 1911. At the Calvary Hill Cemetery in Fulton are buried many of the President's relatives, including his great grandfather, Michael Reagan.

The Longest Road?

ACCORDING TO THE JANUARY, 1997 ISSUE of *Popular Science,* the longest highway in the United States is Route 20. It spans 3,370 miles from Boston, Massachusetts, to Newport, Oregon.

Not so fast!

The Lincoln Highway from Times Square in New York to Lincoln Park in San Francisco, spanned 3,389 miles in 1913, with a variety of factors affecting the final tally, which may also be said of U.S. 20, making the two roads close to a virtual tie. Route 20, however, lacks the romance of Route 66 (2,400 miles from Chicago to Los Angeles, California) and the nostalgia, history, personality and ethos of the Lincoln Highway, a uniquely bonded strip of history, one-of-a-kind in the annals of human record.

The Importance of Good Roads

IN THE EARLY DAYS OF THE LINCOLN HIGHWAY, the federal government showed little interest in promoting roads, choosing rather to leave distance travel to the railroads and local roads to free enterprise. At the turn of the century, who could have predicted America's love affair with the auto—"Machines" as they were sometimes called?

Duryea, Haynes, and those immediately following put auto manufacturing on a roll and nothing could stop it. Except for the computer chip, has any invention ever so completely changed society? Airplanes put transportation into fast forward, but the auto has reached into every wallet, priority, and lifestyle. "Wheels" have become a right of passage for teen-agers, status for the wealthy, investment credit for the salesman, extension of personality for the socially deprived, a last hurrah for the retired male who always wanted a Corvette, a fantasy for the sports fan, wheels for the soccer mom, a deadly missile for the intoxicated and the last freedom for the elderly. Fresh fruit, vegetables and commodities for consumers via trucks, make them an economic necessity for everyone.

In 1900 auto manufacturers in the United States had sold about 8,000 vehicles. One hundred years later, in the year 2000, auto registrations exceeded 217 million with 190 million licensed drivers – and still counting. Without good roads - -?

At the turn of the twentieth century the auto received a mixed review:

W. W. Townsend said in *Motor Age* in 1901 that "The speedy extinction of the horse is popularly anticipated. I do not take this view. He may be relegated to comparative obscurity, and possibly, in course of time to the zoo; but it is not we who shall live to see his extinction."

Charles Duryea wrote in 1909: "The novelty of owning an automobile has largely worn off. The neighbors have one of their own. The whole family has become so accustomed to auto riding that some members generally prefer to ride alone or remain behind while others go."

Woodrow Wilson, while president of Princeton University in 1906 opined, "nothing has spread socialistic feeling more than the use of the automobile—a picture of the arrogance of wealth."

Rural opposition to the auto was waged chiefly at touring motorists. Autos were claimed to be a danger to horses, stock and even crops. Demonstrated opposition ranged from plowing up roads, barb wiring roads, to boycotting car-driving businessmen and even refusing to vote for politicians who owned the nefarious contraptions.

Iowa Department of Transportation

IOWA, And the Lincoln Highway

358 MILES, 1915   50 Lincoln Highway towns, 1915

IOWA IS AN INDIAN NAME meaning "Beautiful Land". This is a wonderful name, chosen not from a name suggestion book, but from the natural observations of the peoples who lived on this land long before it was "discovered" by the white man. This was certainly an appropriate name until Henry Ford and others began appealing to human fantasies. That appeal, taken up by all the relatives of the advertising clan, has today reached a crescendo penetrating all the senses enroute to the wallet. One result is cars, and cars needed paved roads.

Iowa is still beautiful and appropriately named but takes on a different perspective through the "eyes" of a horseless carriage. Unfortunately, Iowa was famous, or infamous, for its mud. It's great for corn but was frustration, aching backs, hot tempers, strange words, ruined expensive boots, exasperation,

Iowa State Hioghway Commision

The Joys of Motoring, 1915

fractured patience, and a challenge to those people determined to drive in spite of adverse conditions. In summer, dry mud became dust everywhere—in the car, the house, on trees and bushes and even on the corn growing within a few yards of the roadway. Mud was also the fire built under politicians to create better road surfaces. However, the cautious approach to paved roads in Iowa was not a twentieth century phenomenon. Its roots were planted deep in Iowa history. Albert Lea says in a little book printed in 1836: "The natural surface of the ground is the only road yet to be found in the Iowa district and such is the nature of the soil in dry weather, and we need no other. The country being so open and free from mountains, artificial roads are little required. A few trees taken out of the way where routes are much traveled and a few bridges over the deeper creeks is all that is necessary to give one a good road in any direction." Seventy years later the proliferation of automobiles demanded a reality check. No longer were hard surfaces the exclusive domain of city streets. They would also cover rural roads.

Automobile registrations in Iowa rose from 799 in 1905 to 147,078 in 1915 and the increased usage raised the demand for hard surfaced roads, with macadam considered the ideal surface. Macadam at that time was not "black top" as we know it today. Macadam was a packed layer of large stones with an added layer of smaller stones compacted into a hard surface, first used in 1795 for the intercity paved pre-Lincoln Highway road between Philadelphia and Lancaster, Pennsylvania. It performed very well for wagons and buggies but automobiles loosened the stones and destroyed the surface, creating untenable maintenance costs. However, gravel proved a practical solution for lesser traveled roads in Iowa because of the generous supply of local gravel. Nevertheless, the state came to the conclusion (temporarily) that there was no one-size-fits-all road surface. Traffic, soil, and financial conditions were the decisive factors in the choice of permanent surfacing. Often the choice came down to maintenance for its length of life, with funding by taxes or bonds with specific guidelines for each.

Concrete was first used for highway surfaces in 1904 but by 1912 it began to achieve some prominence. A mile was built in 1913 west of Mason City, extended into that community in 1915, and later between Mason City and Clear Lake creating the first interurban paved highway in Iowa. These roads, together with the 1918 "Seedling Mile" of the Lincoln Highway in Linn County, proved their ability to withstand heavy traffic and adverse weather conditions. The problem was the construction cost of $30,000 per mile. However, over time low maintenance costs overrode all other options, with

Carillon Tower, Jefferson

A gift to the people of Greene Count by Mr. & Mrs. W. F. Mahanay October 16, 1966. Note the Lincoln Highway post.

concrete becoming the surface of choice.

Enter the Lincoln Highway. The vision of Carl Fisher to build the "Coast To Coast Rock Highway" and to finance it with subscriptions and annual dues got the attention of the cement industry which agreed to support the project on the same basis as auto manufacturers: one-third of one percent of their annual gross for three years, estimated to provide 2.3 million barrels of cement. By the late teens the federal government began to recognize its responsibility for good roads and in 1919 provided some of the cost to pave Iowa roads. This resulted in much of the Iowa Lincoln Highway being surfaced in concrete by 1924. In 1925, with the advent of the federal numbering system, the Lincoln Highway became U.S. 30. In the 1950s federal highways were being widened and straightened, thus often bypassing small towns. This process, though an improvement in safety and efficiency, resulted in isolating many sections of the original Lincoln. Today, much of this nostalgic road remains a paved state or county road, providing delightful byway excursions off the high-speed routes into the charm of the old Lincoln Highway towns and communities.

Bob Owens stands at the entrance of the Greene County Lincoln Highway Center on Lincoln Way in Jefferson, Iowa..

Famous Lincoln Highway Bridge, Tama, Iowa

Built in 1915. My father crossed this bridge on September 2, 1915 on his way from Belle Plaine to Scranton.

IOWA POSTS

Abbe Creek School

Abbe Creek School post near the Lincoln Highway "Seedling Mile" four miles west of Mount Vernon, Iowa.

coutesy of Nova Dannels

Jefferson, Iowa

Front of Bell Tower on East Lincoln Way and Wilson Ave., Jefferson, Iowa.

Belle Plaine, Iowa

Corner of Thirteenth and Fourth in Belle Plaine, Iowa.

Jefferson, Iowa

Private residence, 507 Edgewood, Jefferson, Iowa.

Grand Junction, Iowa

Guardian of the City Hall, Grand Junction, Iowa, on the Lincoln Highway (Main Street).

Scranton, Iowa

Main Street (Lincoln Highway) and Irving in Gazebo Park, Scranton, Iowa.

Jefferson, Iowa

In front of 403 W. Lincoln Way, Jefferson, Iowa.

Carroll, Iowa

In front of the Carroll County Historical Museum.

IOWA POSTS

Carroll, Iowa

SE corner of 17th and Quint on driveway near back of the house, Carroll, Iowa. Note the depth notch near the bottom of the post.

Logan, Iowa

Northeast corner of East 7th and North 4th, In Logan, Iowa.

Dunlap, Iowa

1/2 block south of Main Street on the east side of Route 30.

Missouri Valley, Iowa

North side of alley east of 6th Street, Missouri Valley, Iowa.

Woodbine, Iowa

Southeast corner of Lincoln Way and Third, Woodbine, Iowa

Missouri Valley, Iowa

Historical Museum Center, Missouri Valley, Iowa.

Denison, Iowa

The Park Motel in Denison is on the Lincoln Highway and in the National Register. Inquiring at the desk about possible Lincoln Highway posts in town, I received a "I-don't-know-of-any" reply. On the way back to the parking lot I spied the above post head reclining in a doorway. Another inquiry at the desk produced: "It is waiting to be repaired."

Cedar Rapids History Center

CEDAR RAPIDS, *Iowa*

IN HIS OWN WORDS:

Wednesday, August 25, 1915

Fair. Left Fulton, Ill. Crossed Mississippi River and traveled to Cedar Rapids, Iowa. 90 miles. Roads good.

CEDAR RAPIDS SHARES SOME COMMON themes with most other towns along the Lincoln Highway – Indians, treaties, log cabins, taverns and certain nondescript, stalwart, non-conformist pioneers. For Cedar Rapids it was Sac Chief Black Hawk who lost the Bad Axe River battle which claimed the lives of 1,500 Indians in 1832, and led to the signing of the Black Hawk Treaty. In successive treaties in 1837 and 1842 the Indians gave up their rights and land in all of Iowa. Most Indians scattered to Kansas or Oklahoma but a few Mesquakies purchased eighty acres near the present Lincoln Highway town of Tama—the only remaining Indian settlement in Iowa.

Following close on the heels of these events, in 1837 the first white man arrived at the future site of Cedar Rapids. Osgood Shepherd is characterized by Ernie Danek as a "frontier hunter and trapper by occupation and a squatter by inclination. An opportunist who took advantage of all situations for short term gain. Shepherd consorted regularly with undesirable characters, including horse thieves and counterfeiters." Shepherd built his log cabin (later known as Shepherd's Tavern), on what is now the northwest corner of First Avenue and First Street, along the Cedar River overlooking the rapids. Its location made it a convenient stopping place for new settlers or travelers with suspect agendas. In 1841 Shepherd sold his land and left for an unknown destination. Osgood Shepherd is important because his land became the future site of Cedar Rapids. The new owners surveyed and plotted the land for a town which fronted on the Cedar River and at this site was constructed First Avenue, which was also the Lincoln Highway when my father drove through Cedar Rapids on August 25, 1915.

During the time of Shepherd's Tavern the site was known as Columbus. It was plotted under the name of Rapid City but incorporated as Cedar Rapids on January 15, 1849.

The real genius of Cedar Rapids was not those hardy souls who braved the wilderness to build the first log cabin, nor those with a gift for finance or business or politics. It was the ordinary people—the people who grew up on farms, who had ordinary jobs and who lived ordinary lives—conventional people who demonstrated dependability, honesty and commitment to proven values, who always saw light at the end of the tunnel, whose glass was always half full, who would go the extra mile and would greet adversity with hope for better to come.

Nova Dannels is one of those ordinary people, a child of rural Iowa. She was born and reared on a farm along the Lincoln Highway, 25 miles west of Cedar Rapids - a farm bought by her father, George Meyer, from his father, Christian, in 1915, the year my father drove the dirt Lincoln Highway past this farm on his way to San Francisco.

Nova grew up only six miles west of Youngville, a famous two-story gas station/restaurant built in 1931 by Joe Young. It has been fully restored with period furnishings including a small museum and gift shop. In its heyday it was the only all-night stop between Cedar

Linn County Historical Society

Maxen Electric, 1904, 1913

Built in Cedar Rapids by Roy McCartney, who was serious about his car but could not secure necessary financing. However, he did produce three cars in 1909. One was destroyed by a streetcar, with the fate of the second one unknown. The third resides at the History Center in Cedar Rapids. In 1913 another unsuccessful effort was made to produce the car.

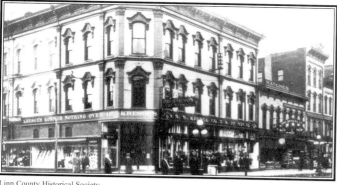

Linn County Historical Society

On The Lincoln Highway, c. 1915 *A view of history on First Avenue in Cedar Rapids. Many years later Kresge esrablished K-Mart*

Rapids and Tama. Nova now lives in Swisher, a suburb of Cedar Rapids, and since 1969 has been a charter member of the History Center Volunteers. In 1928 she watched the paving of the Lincoln Highway past her farm home. She remembers the contractor asking her father for permission to use a portion of the field along the road in which to keep his mules. Nova says, "We ended up with forty mules in that hayfield all summer. We didn't need an alarm clock, since a 40-mule chorus woke us up and the neighbors across the road at 4:00 A.M. Much of the work was done with mules and slip scoops." Growing up on the Lincoln Highway, she was well acquainted with its mud and dust. She reminisces, "Dust all summer, mud in the spring and fall and in the winter mostly snow. Our car usually spent the entire winter in the machine shed."

Nova tells the story of a Lincoln Highway Christmas in 1926. It was an eastern Iowa white Christmas with serendipity guests. Christmas eve, 1926, gave new meaning to "White Christmas." When snow began falling in the morning, the children were happy because Santa would have snow for his sleigh. The snow came faster and faster with increasing wind as the day progressed, so her father went to the barn early to get the cows in out of

Nova Dannels

The Meyer Farm *One of the three Meyer farms on the Lincoln Highway west of Cedar Rapids. This was the host for the snow-bound Christmas in 1926 and was Nova Dannels' home.*

the storm and have an early start on the evening chores.

"About 2:30 there was a knock on the door" said Nova, "and a man all covered with snow said he was stuck in a snow drift out on the road and wanted to know if we could help him to get out. Mom sent him to the barn to find Dad and a few minutes later we saw Dad drive out the lane with Bill and Florie, our big Belgians, hitched to the bob sled. Pretty soon they came back towing a car which he pulled over by the machine shed and the man and his wife came into the house. Dad had no more than unhitched the horses when another man appeared. He was stuck in the same snowdrift. This went on until there were four cars by the shed and four couples in the house. Once more there was a knock on the door and Dad went to get the horses again. This time it was a couple with three children, and by now our house was getting a bit crowded and Mom was beginning to wonder how to feed ten grownups and three children and our own family of five."

Linn County Historical Society

Lincoln Highway Repair, *1915 Lincoln Highway repair on 7th Avenue in Marion, Iowa, near Cedar Rapids.*

Nova Dannels

Lincoln Highway Traffic, 1928

Dwayne Meyer, 7, and his older brother, Everett, 9, drive on the Lincoln Highway in front of their farm home twenty-five miles west of Cedar Rapids. The car was made by their Uncle Lee. Nova Dannels was their 11-year-old sister

As it turned out, all thirteen of the Lincoln Highway travelers were on their way to Cedar Rapids for Christmas, and it was obvious that no one was going anywhere that Christmas Eve. With typical Iowan flexibility mixed with the spirit of Christmas, what could have been a complete fizzle turned out to be a delightful Christmas Eve like no other. At bedtime the guests brought blankets from their cars, along with food they had brought along for Christmas dinner the next day.

Nova describes their evening together: "After supper everyone gathered around our Christmas tree which stood in a corner of our living room with candles ready for lighting. For a few minutes, everyone watched and admired. I can still smell the candle wax and the pine scent from the evergreen tree fresh from the corner of our orchard where Dad had planted them when I was a baby, especially for Christmases to come. After the candles were snuffed out, Mom got out all our extra quilts and blankets and sleeping arrangements were discussed. The six children rolled up in heavy quilts on the living room floor. Two of the young couples fixed makeshift beds on the dining room floor and the others went to our three bedrooms upstairs and the spare room on the first floor. Dad went to the basement to put more wood and coal in the furnace and after that the house was very quiet."

The next day was Christmas. And Nova says, "The sun came up bright and clear and the snow sparkled as if it had been sprinkled with diamonds. The women had dinner cooking by the time the men finished the chores and we children were having a ball with our presents. I don't remember who got what except [for] my beautiful doll. Her name was Dorothy Dimples. It was written on a tag pinned to her dress. She had a beautiful wardrobe made by Grandma Kruger which included a plush coat just like the one she made for me."

Nova said they had just finished eating when her Uncle John, who lived a mile west, called and said that three cars had arrived at his place from Belle Plaine but the snow was too deep to go any further. The dilemma was solved by Nova's father who hitched up the horses to the bobsled and made several trips transporting their guests to her Uncle John's place so they could drive back to Belle Plaine. From there they took the train home.

About the stranded cars, Nova said, "Those five cars sat beside our machine shed most of the winter. Seemed like that snow would never melt, but eventually it did and one by one they got driven away."

BELLE PLAINE, *Iowa*

IN HIS OWN WORDS:

Thursday, August 26, 1915

Fair. Broke camp at 6:a.m. Left for Belle Plaine, Iowa. Booked for 30 to 31st of August. Trip was 37 miles. Good roads.

Friday, August 27

Fair Shot business houses and some kids for show also 20 farms shots. 3 ? miles out.

Saturday, August 28

Fair. Made 60 kids shots for Tuesday show.

Sunday, August 29

Cold. Run off slides and proofs of everything. Cold raw day.

Monday, August 30

Frost, Fair. Shot farmers for show Tuesday, 20 different. Show $36.70, 50% 50%.

Tuesday, August 31

Fair, Passed proofs and run off orders. Show, receipts $33.70, 50% 50%.

Wednesday, September, 1

Cool. Delivered up everything in Belle Plaine, Ia. Delivery $51.75.

Thursday, September 2

Left Belle Plaine 6 A.M.

IN AUGUST 1915 MY FATHER SPENT EIGHT days in Belle Plaine photographing children, farms and businesses. These must have been a busy eight days because he "shot" over twenty children plus forty farms and local businesses. He booked the local theater for two nights to show off the children, and local business enterprises.

Eighty-five years later my wife and I visited Belle Plaine as we followed my father's diary. Many of the buildings we passed were no doubt the same ones he saw in 1915. Is it not reasonable to believe that in some attic, some family album somewhere, would still repose some of those 85-year-old images? We ate breakfast at the Lincoln Café, "shot" the Lincoln Highway concrete post along with George Preston's famous gas station close by, and then enjoyed the privilege of visiting Bev and Wallace Winkie, retired long-term Belle Plaine school teachers who came to teach and stayed. They in turn introduced us to Don Villers, historian, and Don Magdefrau, publisher of the *Belle Plaine Union*. If all the people of Belle Plaine and its environs are as gracious as these folks, it is not hard to understand why people enjoy living here. In spite of all those combined efforts we were unable to blow the dust from any 85-year-old prints made by my father.

However, all was not lost. The Winkies shared some interesting Belle Plaine history with us. Apparently word of the fertile soil in Iowa drifted south as far as Greene County, Tennessee where Hyrcanus Guinn, a friend of Andrew Johnson's, lived. He with his wife and brother came to Belle Plaine to check it out. They returned to Tennessee for the inauguration of Andrew Johnson as governor in 1852 and then returned to Belle Plaine followed by other family members, the Ealys and Greenlees. These were the years of pioneer development of many would-be Lincoln Highway towns, largely

George Preston's Legacy

George Preston Station, Thirteenth and Fourth in Belle Plaine, Iowa

Lincoln Cafe

Thirteenth and Sixth Avenue in Belle Plaine

motivated by the arrival of railroads. For Belle Plaine it was the Chicago and Northwestern in 1862.

Fifty-one years later in 1913 it was the Lincoln Highway which motivated development. The prospect of Belle Plaine being on this famous cross-country route was an occasion for genuine excitement in this small eastern Iowa farming town. First the trains brought goods, markets and transportation, then the Lincoln brought customers, a fact not unnoticed by the people of Belle Plaine, and they were equal to the challenge. The Lincoln Café, George Preston's station, the Herring Hotel, the F. L. Sancot Garage and many other businesses prospered from that 1913 decision to aim the Lincoln Highway at Belle Plaine.

Unfortunately, in 1936 someone noticed that a straight line is the shortest distance between two points. In this case the straight line was from Cedar Rapids to Tama, adding Belle Plaine to the growing list of Lincoln Highway towns known as "by-ways." Growing concerns for rapid travel and safer highways was becoming a priority for those persons responsible for such things so Belle Plaine is now about six miles south of Route 30.

Highway Renovation

The Lincoln Highway in Belle Plaine, Iowa, receives some needed attention.

Belle Plaine

Corner of Thirteenth and Fourth in Belle Plaine, Iowa

SCRANTON, *Iowa*

IN HIS OWN WORDS:

Thursday, September 2, 1915

Traveled all day [from Belle Plaine] made Scranton, Ia, 146 miles. Camped between Scranton and Giddem, Ia.

[He probably meant "Glidden"]

SCRANTON, IOWA, IS ONE OF THOSE TOWNS bigger than the sum of its parts. Its history is linked to the history of Greene County, organized in 1854 with a population of 150, with Jefferson as the county seat. The county was named for Gen. Nathaniel Greene, a hero of the Revolutionary War.

The first courthouse was the log cabin of Judge William Phillips. A true courthouse was built two years later, in 1856, and a newer one fourteen years later in 1870. This one is the one my father would have seen, if he had had an occasion to look for one on September 2, 1915. The present one was built two years later in 1917 for $179,752.66.

Greene County seemed to have an attraction for presidential candidates witnessed by the fact that candidate Harold Stassen launched his campaign from the south balcony of the Greene County Courthouse in 1948, and in 1952 Dwight and Mamie Eisenhower made a whistle stop at the Jefferson depot. This, however, was not the first time Dwight Eisenhower visited Greene County. He was a young lieutenant with the 1919 army convoy which traveled through Scranton on the Lincoln Highway en route to San Francisco.

Scranton is not without its own personal vignettes. There is the intriguing case of the "accidental" driveway. As the story goes, James Holden, one of the three original Iowa State Commissioners, lived in a large farm home (now abandoned) just east of the Lincoln Highway. The Lincoln was heading south on Main Street but at some point must turn east. If the route continued straight south two more blocks, beyond the Holden road, it would eliminate two additional turns. However, when the final

Typical Iowa Image

Grain elevators are a common scene in the corn/wheat country of Iowa.

93

Scranton

Lincoln Highway, Scranton, Iowa

route was chosen and the concrete paving all finished, guess whose house the Lincoln Highway passed by and who sported the only concrete driveway in town?

Gregory Franzwa quotes a heartwarming story from the September 4, 1919, *Scranton Journal:*

"About two weeks ago W. O. Schuyler met a small boy on the Lincoln Highway who was headed on foot for Peoria, Ill. The boy asked Mr. Schuyler for a ride in his car and Walt brought him to Scranton. The lad gave his name as Clifford Maple and said he had left Peoria with his mother's consent and had accompanied some traveling horse traders as far as Glidden, where he deserted them and started for his home because he said they had misused him. Mr. Schuyler took him to his home for a night's lodging, fed him up, and the next day took up a collection and bought him some clothes and a ticket home besides sending the lad away with a small purse of money. Mr. and Mrs. Schuyler were very kind to this lad and have since received a letter from the boy's mother bringing news of his safe arrival home and expressing her heartfelt thanks and gratitude."

Scranton

Main Street (Lincoln Highway) and Irving in Gazebo Park, Scranton, Iowa.

CARROLL, *Iowa*

IN HIS OWN WORDS:
(heading back east)

Wednesday, December 15, 1915

Arrived and got located in Carroll, Ia in Brees Bldg. N. 5th St. Had 1/2 ft. of snow and sleet to \contend with. Also drifts.

Thursday, December 16, 1915

Very cold.. 8 below zero. Left Carroll 6:35 for Vail. Booked Vail for Monday night. Slides and Victory.

Friday, December 17, 1915

Cold snowy. Left 7:10 for Ralston & Scranton. Shot these towns 9 flashes. Got back (Carroll) at 5:00 P.M.

Saturday, December 18, 1915

18 below zero. Run off proofs and slides of Vail and other town.

Sunday, December 19, 1915

Very cold. In town (Carroll) all day. Got slides ready for show.

Monday, December 20, 1915

Fair. Left 6:35 for West Side. Passed proofs and went to Vail. Had a $14.00 house. Also passed proofs in Vail.

Tuesday, December 21, 1915

Left Vail 7:11 for Denison to see if town is worked. Returned to Carroll and run off Vail and West Side delivery for del. Wed.

Wednesday, December 22, 1915

Cold. Left 6:35 for West Side. Delivered—and also in Vail and returned to Carroll at 6:30. PM. Good delivery.

Thursday, December 23, 1915

Fair. Moderate. Left 6:35 A.M. for Manning, Iowa. Town was just worked. Returned to Carroll.

Friday, December 24, 1915

Cloudy. Left town 6:35 A.M. Shot up Arcadia, Ia and Maple River. And booked slides and Victory at Breda.

Saturday, December 25, 1915

In town all day. (Carroll)

Sunday, December 26, 1915

Very cold. In doors all day. Made up delivery and duplicates for Vail.

Monday, December 27, 1915

Very cold. Left 7:50 A.M. Shot Wall Lake and Breda. 45 shots all told for week. Booked in Breda $29.00

Tuesday, December 28, 1915

Cold. Went and shot up Lohrville, Ia. Booked for Monday Jan 3, 1916. Made 28 shots.

Wednesday, December 29, 1915

Clear and cold. Passed proofs in Wall Lake. Show in Breda went $19.60. Also Victory.

Thursday, December 30, 1915

Snow, cold. Passed proofs in Breda, Ia and got back to Carroll 4:00 P.M..

Friday, December 31, 1915

Warm and rain. Delivered in Breda and Wall Lake. Good delivery. $50.00 and some duplicates. Saloon went out of Iowa this day. Made $11.00 with Saloon man in Breda, Ia the day he closed. 5 dozen pictures.

CARROLL, *Iowa*

ITS A SAFE BET THAT EVERY TOWN has had its share of tragedies, some greater, some lesser, but September 25, 1879 was burned indelibly into the history of Carroll. It was about 4:00 A.M. when Engineer Frank Crow, braking his west bound freight to take on water, noticed flames in Henry Schappman's saloon on the north side of Fourth Street. He sounded the train whistle with a continuous blast rousing the citizens from their sleep. In a few minutes they witnessed their worst nightmare. The city's $80 wooden buckets and three ladders were far from adequate for the task.

The sun-dried all-wood structures burned at will, fanned by a strong breeze from the south. By 6 A.M. the fire began to abate because it had consumed everything in a two-block area of the business district. Ninety percent of the town's businesses were now charred ruins. The estimated loss was $200,000, with only $35,000 to $40,000 covered by insurance.

An ad in the *Carroll Herald* shortly after the fire reflects some of the pain and severe loss:

"I wish to say to those in any way indebted to me that in the late Carroll fire, I was a heavy loser. My place

James Kerwin

Community Lifeline

Prior to the Lincoln Highway and autos, trains were the source of people, merchandise, news and markets for the pioneer towns. For Carroll the railroad was the midwife in its birthing. The very land Carroll was built on was bought from the railroad.

of business, my house, nearly all my furniture and goods were destroyed, my loss is upwards of $5,000.00 and if ever I was in need of money, it is now. Do not wait to receive notice from me but come and pay quickly."

Signed: F. E. Dennett

Carroll was severely injured, the fire destroying the very center of its economic infrastructure, but, as we would expect, the town did rebuild. The new, strong building codes required that all buildings in the business district to be constructed with outer walls of brick and mortar and roofs of metal or slate. One year later a large Babcock fire extinguisher on wheels arrived, and in 1882 (the year my father was born) the fire department received a new hook and ladder truck from the G. M. Needles Company of Atlanta, Iowa. Ironically, most of these replacement buildings later fell victim to Twentieth Century urban renewal and were torn down for more modern versions.

In September of 1915 when my father drove through Iowa on his way to San Francisco, he made no mention of Carroll, but on his way back in December, Carroll became the center of his western Iowa activity. On December 15, 1915, he settled in at the Brees building on North Fifth Street. The B. H. Brees building, in its earlier life, was a saloon and became one of the casualties of the 1879 fire. No record could be found of its rebirth but in some fashion at least it gained new life because my father found refuge there and apparently accommodations sufficient to be the center of his photographic operations for the next seventeen days or perhaps longer. His diary ends on December 31, 1915 and he is still in Carroll with at least one theater showing in Lohrville for January 3, 1916.

He did not patronize hotels and rooming houses very often, preferring to camp in his truck. However, the Iowa weather may have had some influence on his decision that fifteenth day of December. He was greeted in Carroll by six inches of drifting snow and sleet. The next morning the temperature was eight below zero and dropped to eighteen below by December 18. I suspect the Brees building on North Fifth Street looked pretty good to him.

With Carroll as command post he ventured into neighboring towns and communities photographing children, businesses and farms. On the sixteenth of December he went to Vail where he booked a theater for December 20. On the seventeenth he worked Ralston and Scranton. The day before Christmas he worked Arcadia and Maple River.

He spent Christmas in Carroll.

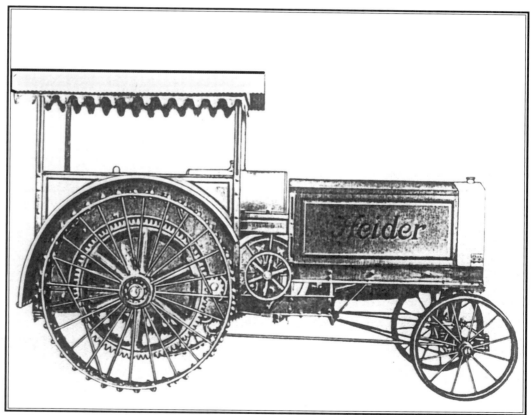

Ready For Field Work

1915 Heider tractor Model C. This tractor was restored by Omer Langenfeld, with over 34 years with the Heider Company, and Richard Collison. It is on display in the agricultural building at Swan Lake Park.

James Kerwin

Carroll Straight Ahead

Lincoln Highway approaching Carroll from the west.

Apparently, his work made a good impression on a Breda saloon keeper, who ordered five-dozen pictures. Since all breweries in Iowa were closed by law as of January 1, 1916, this saloon-keeper evidently wanted some pictorial evidence of his enterprise for posterity. During those seventeen days between December 15 and December 31, my father fanned out from Carroll to twelve surrounding towns. Zero and below temperatures makes it a story in itself, but doing so in an open, two-cylinder truck in snow makes it even more remarkable. The bright side is that the roads were frozen. Had it been springtime, I wonder which mud hole would have done him in.

Seeds for the city of Carroll were planted in 1856 when the general assembly made provision for a new county in western Iowa. It was named Carroll in honor of Charles Carroll, a man from Maryland, with noteworthy credentials. The wealthiest man in the colonies, he was one of the fifty-six signers of the Declaration Of Independence and outlived them all. Carroll was the only one of the signers to attach the location of his home to his signature so that King George would know which Charles Carroll it was who declared his independence.

In 1867-68 the Cedar Rapids-Missouri Railroad (later Chicago Northwestern) built through Carroll County and selected sites for towns along the way. The federal government, to help the railroads pay for laying the track, gave them land twenty miles on either side of the tracks to sell. In August, 1867, the railroad plotted the land which is now Carroll. That fall Carroll also became the county seat and was incorporated in 1869 with I. N. Griffith, who also operated a general merchandise store, as mayor. The first family was that of A. L. Kidder who became postmaster and operated a grocery store and restaurant. By now the population of Carroll was 384. Ten years later, at the time of Carroll's great fire, the population had grown to 1,200. The Lincoln Highway made its official appearance in 1913, and in 1915, when my father came to town, the population was over 4,000. Today, Carroll's population is stabilizing at around 10,000. It is in the heart of a rich farming area and enjoys the amenities of the Middle Raccoon River. With I-80 some forty-five miles to the south, Carroll has become one of those must-see Lincoln Highway byways.

17th and Quint

SE corner of 17th and Quint, on driveway near back of the house, Carroll, Iowa. Note the depth notch near the bottom of the post

Carroll

In front of the Carroll County Historical Museum steps

ARION / DOW CITY, Iowa

IN HIS OWN WORDS:

Friday, December 10, 1915

Cloudy and snow. Run to Dow City. Booked for Monday.
Shot schools and stores in Arion, Ia. Snow 1/2 ft. deep.

Saturday, December 11, 1915

Cold. Shot up some business houses for show and some stores.

Sunday, December 12, 1915

Very cold. In town all day. Run off slides and proofs.

Monday, December 13, 1915

Cold. Run off P. I. for druggist and passed Arion proofs. Show rec. $23.00.

Tuesday, December 14, 1915

Very cold. 6 below zero. Delivered and passed proofs in Arion and Dow City and moved truck back to Arion at hotel in Arion.

Wednesday, December 15, 1915

Very cold, stormy, 1/2 ft. of snow. Left Arion 7:30 A.M.

MANY TOWNS ALONG THE LINCOLN Highway share a common history with the railroads affecting not only their economic and social development, but transportation as well. It is a monotonous story. The evolution of transportation in the United States began with footpaths beaten through the wilderness by Indians in their search for game, or on their seasonal migrations. Then came horses, utilitarian wagons, and genteel buggies. These in time gave way to parallel steel rails stretching from one coast to the other. Towns were built along these rails and with the birth of the auto and its multiplied siblings, these paths and wagon trails became car-wide mud holes as long as the rain storm. In spite of the mud, these roads gradually displaced the railroads in moving people, merchandise, prosperity and diverse cultures from coast to coast.

Towns were eager, and in some cases even fought, to have the Lincoln Highway come to town. For a town to

Arion Post Office today

Cindy Simon

Arion Depot

Arion depot, c. 1912. The decision for the Arion depot was made by the state General Assembly over the objections of nearby Dow City and Belltown who feared business competition. This is how my father would have seen the depot in 1915.

Arion Bank

Arion State Bank was organized in August 1902 with capital stock of $10,000. Today it is vacant and stands across the corner from the Green's Hotel lot.

be on the famous Lincoln Highway was not only prestigious but a boon to the local economy. Unfortunately, the mud holes were replaced with traffic jams, which in turn gave birth to the bypass, removing traffic from downtown and along with it sales and commerce. But the insatiable appetite for cars bred still more bypasses around the bypass. They are called Interstates. Many towns in order to survive have moved their commerce to the exits.

Arion, Iowa, is an exception to most of the above except for the role of the railroads. Arion was born out of controversy resolved by the General Assembly. It did not begin with one railroad. It began with the juncture of three railroads: Northwestern (1866), Milwaukee (1887), and Illinois Central (1899).

The debate centered on the proposal to establish a depot at the crossing of the Northwestern and Milwaukee lines. The neighboring towns of Belltown and Dow City claimed it was too close and would hurt their business. Denison, however, favored a depot to facilitate the trans-

The Arion Railroad Tower

The Arion Railroad tower, c. 1912, was a long-time landmark for Lincoln Highway travelers. It was built in 1887 when the Milwaukee line crossed the Northwestern. It was razed in January 1969.

fer of passengers to Denison. The General Assembly in February 1888, chose to accommodate the traveling public rather than local interests by voting to establish Arion at the juncture of these three railroads.

Arion was first named Lydia after Lydia Lanning, who came to teach school in 1886 and became the first postmaster in 1888. Two and a half years later in 1890 the name was changed to Arion. The sale of lots became a hot item in May 1888, selling from $25 to $70 each. All lots were soon sold and Arion was incorporated in 1894. The railroads brought prosperity but also enough lawlessness to get the attention of historians. "Due to the three railroads, Arion's early history was marked by some lawlessness on account of the fact that it was a convenient stopping point for that class of undesirable citizens known as tramps."

Three hotels in Arion did a thriving business serving the three railroads. It was a booming center of commerce boasting a general store, meat markets, grocery and

Greens Hotel

My father stayed in the Green's Hotel, December 14, 1915, on this lot. The hotel burned down in 1921 and was not rebuilt.

hardware stores, a plumbing establishment, harness shops, a shoe and furnishing goods store, barbershop, restaurant and lunchroom, pool hall, saloon, lumberyard, livery stable, blacksmith and wagon shop, two elevators and grain businesses, bank, garage, hospital, doctor's office and drugstore, building contractor, mason and dray wagon, telephone office, and three railroad stockyards.

Arion had its good times but also its tragedies. There were at least five devastating fires over the years:

1905 – The first county home.

1909 – Arion Merchantile along with several adjoining businesses.

1921 – Green Hotel along with an entire row of businesses. These were never replaced.

1920s – The "Square Hotel," along with the Illinois Central depot. These were never rebuilt.

1985 – The Congregational Church parsonage and library.

When my wife Miriam and I visited Arion on June 8, 2001, we saw none of its earlier grandeur. We saw the present small post office, the empty, hollow-eyed Arion State Bank building, the vacant lot where once stood the Greens' Hotel (the most likely one my father patronized), and the former two-story yellow brick school building, now in commercial use. We also visited a resident of Arion, who shared some local history. He told about the annual "Buffalo Day" in September which began in 1907 and eventually became the Crawford County Fair. Schools throughout the county were closed so students could attend. Don said buggies were parked everywhere. It was a gala event with such attractions as cooking contests, dances, and livestock displays. The fair was finally moved to Denison. Only the Illinois Central and the Chicago and Northwestern (now the Union Pacific) railroads remain today. By 1941 the railroads abandoned their respective depots and in 1969 even the railroad tower, long a landmark for the Lincoln Highway, was razed.

My father did not even mention Arion on his way west. He drove through it on September 4, 1915, stopping at Dunlap, where he met "parties on foot from Philadelphia," and camped for the night. The next day he took them along to Omaha. It was on his way back east on December 10, 1915, that he and Arion became acquainted. The weather was cold with six inches of snow as he photographed children and businesses in both Arion and Dow City. Apparently there was no theater and it was too cold to set up out of doors like the community did for Buffalo Days. Perhaps one of the hotels provided the necessary facilities. One wonders why he passed by a flourishing city, like Denison, for a small railroad town. Perhaps it was his rural, down-on-the-farm background that made him more comfortable in smaller communities. He doesn't say how many photos he took.

But I still wonder, as before, if some still exist in dusty family albums. In fact, some of the unidentified period photos and negatives preserved in our family photo boxes may have been made in Arion or Dow City.

DUNLAP, *Iowa*

IN HIS OWN WORDS:

Saturday, September 4, 1915

Fair. Traveled 60 miles to Dunlap, Ia Met parties on foot from Philadelphia.

Sunday. September 5, 1915

Camped with Phila. Parties. Broke camp and took them along to Omaha, Neb.

[December visits to Dunlap were on his way back East]

Thursday, December 2, 1915

[Missouri Valley] left town [Omaha] by myself for Mis Valley. Made 6 shots. Canvasse 350.

Friday, December 3, 1915

Very cold. Delivered in Missouri Valley and left for Woodbine, NG. Took next train to Dunlap. Booked for 6th and 8th slides and Victory.

Saturday, December 4, 1915

Shot up town for show. Got 40 some flashes and all schools. Run into Omaha at night.

Monday, December 6, 1915

Very cold. Ging [an employee] went home. Arrived in Dunlap 4:30 P.M.

Tuesday December 7, 1915

Cold. Made some few shots for show. And run proofs and slides for show. Receipts for show, $30.60.

Wednesday, December 8, 1915

Fair. Made trip into country with truck. Shot farms for show Thursday.

Thursday, December 9, 1915

Run off proofs and made some hurry up orders. Receipts for show $13.00.

IOWA, IN SOME RESPECTS, shares a common ancient history with Illinois. Archeological evidence indicates that this stretch of North America was inhabited 10,000 years B.C., by the Paleo Indians, who stalked the big game of the glaciers. Unique to the area (and China) is a finely ground windblown silt called loess, formed 20,000 to 10,000 years B.C. As the Pleistocene glaciers melted, large areas of flood-deposited sediments were left exposed to wind which deposited them in a long narrow band along the length of the Missouri River Valley creating what is now the Loess Hills and the Loess Hills State Forest. These deposits in some cases go to a depth of 200 feet. The physical properties of loess with its very low shear strength when water-saturated marks Iowa as having one of the highest erosion rates in the nation – averaging forty tons per acre per year. Believe it or not, this soil instability also had its effect on the Lincoln Highway. Imagine yourself trying to maintain those roads with soil that would collapse under your feet as soon as it got wet.

The first Europeans to explore what is now Harrison County were Lewis and Clark on their Missouri River trip in August 1804. The first European settlers were thought to be Mormon dissenters between 1845 and 1847, during the Mormon migration to Salt Lake City. Among early settlers were Daniel Brown in April

Dunlap

Miriam stands beside the Dunlap post, one-half block south of Main Street on east side of Route 30.

Main Street, Dunlap, Iowa, 2001

1847 and Uriah Hawkins, in July 1847; The Barney brothers, John Reynolds, John Harris, Amos Chase and Silas W. Condit came in 1848; Orville Allen and Alonzo Hunt in 1849.

Harrison County was named for the ninth President of the United States, William Henry Harrison. The county was established in 1851 and officially organized March 7, 1853, by an act of the fourth General Assembly of the State of Iowa.

The Harrison County population grew from 1,065 in 1854 to 15,666 in the year 2000. Dunlap, 1,139 population, was named for George L. Dunlap, a railway official—additional evidence of the strong mutual relationship between many Lincoln Highway towns and the railroads. Missouri Valley, 2,992 population, was named after its location in the river valley which consumes one-fifth of Harrison County.

1915 • IMPORTANT EVENTS

- United States President: Woodrow Wilson.
- Albert Einstein developed the Theory of Relativity.
- The British ship *Lusitania* was sunk
- The first transatlantic radiotelephone communication is sent between Arlington, Virginia, and the Eiffe Tower Paris.
- The Boston Red Sox won the World Series.
- Average annual income: $1,267
- Average price of a new car: $390.

- The Panama-Pacific International Exposition in San Francisco.
- The United States Government minted the first $50 gold pieces for the Panama-Pacific International Exposition.
- The taxi industry begins when automobile owners discover that people will pay for a short auto ride. Fare is a "jitney" (a nickel) and cars are called jitneys.
- The Ford plant in Detroit produces its one-millionth automobile.

MISSOURI VALLEY, *Iowa*

IN HIS OWN WORDS:

Thursday, December 2, 1915
*Left town by myself for Missouri Valley.
Made six shots. Canvassed 350.*

Friday. December 3, 1915

*Very cold. Delivered in Missouri Valley
and left for Woodbine. Woodbine NG.*

WITH GRASS AS HIGH AS THE BACK of a horse, land teeming with elk, buffalo and prairie wolves, Lewis and Clark were probably the first explorers to set foot in Missouri Valley. Prior to 1849 not a white settler had set a plow, driven a stake, or set up a house there. Conflicting records confuse the names of the first settlers, but there is a kind of combined agreement that between 1849 and 1851 names like William Smith, William Dakan, John Reynolds, Charles Smith, Adam Smith, George Lawrence and Mongrum were found scattered among the tall prairie grass. By 1853 other settlers were arriving and buying land for $1.25 per acre. The fertile land was good to them, yielding wheat, corn, fat cattle, and wild fruit. They were proud of their new homes and induced family and friends to join them. A description of the character and optimism of these early folk is found in the *Missouri Valley Area Centennial:*

I once knocked at the door of one of these enterprising men late in the evening, a stranger in a strange land, homesick, hungry, tired and weary. Expecting to be told that his house was small, family large, with a few other excuses, and that, "I can't keep you," which to me appeared to be really a fact. Feeling, however, that I could not pass the semblance of a chance to stop without a trial. I meekly addressed the proprietor, asking the privilege to stop with him for the night; to which he said, "Certainly," and seemed pleased to see me. After making me comfortable, and feel that he was my friend, he sat down, and to the best of my knowledge, asked 27 questions in half an hour, and frankly answered as many more for me the next half hour, telling all about the country, choice pieces of land, good chances for speculation, laughing at my notion of being homesick, saying if I stayed here a little while I would be homesick if I left, and this experience has proven true. I thought he was an exception of a man, (and I half believe yet he was,) but this was the general style of the people, hale fellows well met, sociable, accommodating and agreeable.

Missouri Valley is beautifully situated at the foot of the bluffs, one mile from the Boyer River, six miles from the Missouri, and is at the juncture of the Chicago and Northwestern, Sioux City and Pacific, and the Fremont, Elkhorn and Missouri Valley railroads.

In 1854 H.B. Hendricks from Putnam County, Indiana halted his team on the present site of Missouri Valley. In 1856 M.B. McIntosh and his brother, George R. arrived also from Putnam County, Indiana. These people could truly be called the "Town Fathers" of Missouri Valley. By 1867 a bona fide town had emerged. That year alone welcomed the first general merchandise store, a blacksmith, harness maker, physician, attorney, exclusive clothing store, grocery store, stock and grain sales, tailor and post office. The next year, 1868, saw the first newspaper (July 3, 1868), *The Harrisonian*, lumberyard, jeweler, hardware and drug store. The great number of businesses seems strange in only two years. Perhaps the post-Civil War era awakened the pioneer spirit of many families and made seeking ones fortune in the West an enticing option. The population to support these

Russell Rein

A Bird's-eye View of Missouri Valley, Iowa.

enterprises was only 600 but the activity generated by the junction of three railroads, and the resulting potential for growth, was not lost on easterners looking for a lucrative future. In fact, the freight agent, Waldo Abeel, of the Chicago & Northwestern Railroad reported a net income of the railroad in Missouri Valley of $35,000 per month.

One-hundred-thirty-three years later, Missouri Valley, with a population of 3,000, still offers the advantages that attracted so many pioneers long ago. Transportation was then and still is an attractive Missouri Valley asset. In addition to the Union Pacific Railroad, the community boasts two Interstates (I-29 and I-680).

North Side

North side of alley east of 6th Street, Missouri Valley, Iowa.

Iowa State Welcome Center

1915 AUTOS

The following automobiles ("machines") began manufacture in 1915–the year of the Panama-Pacific International Exposition, and the year my father drove from Eastern Pennsylvania on the Lincoln Highway to the Expo in San Francisco. Of the thirty-eight marques, none exist today with only Brewster (1915-1935) coming of age. Average age: four years.

Andover, (1915-1917) Andover, Massachusetts (Electric)

Apple, (1915-1917) Dayton, Ohio $1,150

Bailey-Klapp, (1915) Elwood, Indiana

Bartlett, (1915) Philadelphia, Pennsylvania (electric for taxi service)

Biddle, (1915-1922) Philadelphia, Pennsylvania $2,000 - $4,000

Ballstrom, (1915-1921) Battle Creek, Michigan

Brewster, (1915-1935) Long Island, New York & Springfield, Massachusetts, $3,500 - $8,500

Carnegie, (1915) New York, New York. An auto mail scam. Wm. J. Bailey spent 30 days in jail.

DeKalb, (1915) St. Louis, Missouri

Detroit, (1915-1917) Detroit, Michigan

Dort, (1915-1924) Flint, Michigan $1,000

Driggs-Seabury, (1915-1923) Sharon, Pennsylvania & New Haven, Connecticut $395-$1,975

Eagle Electric, (1915-1916) Detroit, Michigan $1,000 - $1,475

Elco, (1915-1917) Sidney, Ohio $485

Farmack, (1915-1916) Chicago, Illinois $855 - $1,155

Harding Twelve, (1915-1916) Cleveland, Ohio $2,000

Harvard, (1915-1921) New York & Maryland $750 - $850

L.C.E., (1915-1916) Waterloo, Iowa $1,550 - $1,650

Macon, (1915-1916) Macon, Missouri $350 - $400

Madison, (1915-1919) Anderson, Indiana $1,050 - $1,550

Mecca, (1915-1916) New York, New York $450 - $695)

Meteor, (1915-1930) Piqua, Ohio $4,850 - $5,500

Midget, (1915) Springfield, Massachusetts $325

Monitor (1915-1922) Columbus, Ohio $759 - $3,475

M. P.M. (1915) Mount Pleasant, Michigan $1,095

Mulford, (1915-1922) Brooklyn, New York

Niagara Four, (1915-1916) Buffalo, New York $740

Ogren, (1915-1923) Chicago & Waukegan, Illinois $2,500 - $3,900

Peters, (1915) Philadelphia, Pennsylvania $390

Pilgrim, (1915-1918) Detroit, Michigan $685 - $835

Piliod, (1915) Toledo, Ohio $1,485

Royal, (1915) Bridgeport, Connecticut $250

Sterling, (1915-1916) Brockton, Massachusetts $550 $650

Stewart, (1915-1916) Buffalo, New York $1,950

Storms Electric, (1915) Detroit, Michigan $650 - $950

Volta – Car, (1915-1916) New York, New York $585

Waco, (1915-1917) Seattle, Washington $950

Yellow, (1915-1930) Chicago, Illinois & Pontiac, Michigan $1,800 - $6,500 (genesis of Yellow Cab Co.)

NEBRASKA, *And the Lincoln Highway*

DISCOVERY OF NEBRASKA

Now let us climb Nebraska's loftiest mount,
and from its summit view the scene below.
The moon comes like an angel down from heaven;
Its radiant face is the unclouded sun;
Its outspread wings the over-arching sky;
Its voice the charming minstrels of the air;
Its breath the fragrance of the bright wild flowers.
Behold the prairie, broad and grand and free –
'Tis God's own garden, unprofaned by man

"Nebraska. ----A poem," 1854.

Miles: 460 (1924) Towns: 47 (1915)

WHERE THE WEST BEGINS is a matter of opinion. Although Nebraska has the distinction of commanding the Lincoln Highway center of the United States, population density gives the perception of the center being farther East particularly to those of us who have always believed that the migration moved from east to west. A closer look at history, should provide a deeper appreciation of this great land of corn and cattle.

The march of Francisco Vasquez de Coronado with his army of 1,000 men from Mexico to the Platte Valley in Nebraska (1541) was an adventurous undertaking nearly a century before Jean Nicolet established relations with the northern Indians. The Hon. James W. Savage, judge of the Third Judicial Court, gave a speech to the Nebraska State Historical Society on April 16, 1880, in which he revealed the results of his historical research. His Honor claimed "— fourscore years before the Pilgrims landed; sixty-eight years before Hudson discovered the [Hudson] River; sixty-six years before John Smith; twenty-three years before Shakespeare was born; Nebraska was discovered; the peculiarities of her soil and climate noted, her fruits and productions described, and her inhabitants and animals depicted." This is helpful to know but since none of Coronado's 1,000 men stayed, lived in sod houses or raised families, we will still take off our hats to those pioneers from the East who did. Coronado was looking for gold but did not find it. The pioneers of Nebraska were looking for a future and found it.

Our interest in Nebraska, of course, was the little brown leather diary which I found in a box of unmarked photos. The more we read the diaries and scanned maps, the more imminent was a change in travel plans from a land-based trip to Alaska to a diary-based trip over the Lincoln Highway to San Francisco. In fact we did it twice: May-June, 2000 and again in June, 2001.

However, it was meeting the people of Nebraska that turned on the lights! It's dangerous to name names but in this case I dare not resist: Mildred Heath, Overton; Paul and Helen Casper, Elm Creek; Jane Bernhard, Shelton; Ann Anderson, Gothenburg; Bob Stubblefield and Tom Lutzi, co-chairmen of the Lincoln Highway conference in Grand Island; Fae Christensen, Paxton; Betty Freeman, Sidney; plus a multitude of interesting and wonderful people in restaurants, truck stops, small businesses, museums, newspaper offices, court houses – even policemen, were all great ambassadors for the Lincoln Highway and the State of Nebraska which gives the lie to the notion that all Nebraskans were born in a mud hole with hayseeds in their hair.

Elkhorn, Nebraska

Two replica Lincoln Highway posts were dedicated east of Elkhorn, Nebraska, on a stretch of the original Lincoln Highway..

Elkhorn, Nebraska

Southeast corner of Lincoln Highway and 192nd Street in Elkhorn, Nebraska. Note that the replica medallions are missing in all the Elkhorn concrete posts. They get recycled at flea markets.

Elkhorn, Nebraska

Northeast corner of Lincoln Highway and 192nd Street, Elkhorn, Nebraska. Note Lincoln Highway sign on telephone pole, and the UPRR crossing.

Elkhorn, Nebraska

South side of Lincoln Highway near the LH Memorial, Elkhorn, Nebraska

Grand Island, Nebraska

Southeast corner of second and Cleburn, Grand Island, Nebraska

Grand Island, Nebraska

Northeast corner Tilden and Second at the Nebraska Department of Roads, Grand Island, Nebraska.

Shelton, Nebraska

1/2 block west of C street on south side of Lincoln Way, Shelton, Nebraska. Bob Stubblefield's son helped place this post in 1998. It accompanies a time capsule to be opened July 4, 2025.

North Platte, Nerbraska

Corner of Jeffers and 5th Streets in front of North Platte College, former post office.

Kearney, Nebraska

In front of the Nebraska Department of Roads, Kearney, Nebnraska.

North Platte, Nebraska

Southwest corner of Jeffers and 5th Street North Platte, Nebraska

Kearney, Nebraska

Lincoln Highway dedication post at the Great Platte River Road Archway Monument, June 14, 2001, Kearney, Nebraska.

North Platte, Nebraska

"Shorty," southwest corner of 4th and Taft, North Platte, Nebraska

Lexington. Nebraska

Entrance to the Dawson County Museum, Lexington, Nebraska

Paxton, Nebraska

Northwest corner of Spruce and Third, Paxton, Nebraska

Paxton, Nebraska

Southwest corner of Oak and Third, Paxton, Nebraska

Sidney, Nebraska

West of town on the north side at historical marker, Sidney, Nebraska.

Special Collections, University of Michigan Library

Henry Joy

Henry Joy (left) and Austin Bement are joyful, having just extricated their Packard from a Nebraska mud hole on their way to San Francisco in 1915.

Ribbon Cutting, Grand Island

Dedication of the seedling mile at Grand Island, Nebraska, June 16, 2001.

Original Brick

Two mile stretch of original brick Lincoln Highway, adjacent to the UPRR, east of Elkhorn, Nebraska. We ate lunch in the shade of the mulberry trees on the right.

1917 Hudson

Burton Ristine's 1917 Hudson adds a period touch to the 2001 conferees walking the approach to the Gothenburg Platte River Bridge. My father traveled over this berm to cross the bridge on October 2, 1915.

Boot Hill, Ogallala, Nebraska

"The town too tough for Texans." Of Ogallala's fifty buildings, not one church spire pointed upward. Half the businesses were dance halls, gambling houses and saloons. Boot Hill became the final resting place for cowboys, settlers and drifters who came to Ogallala, some for a meaningful future, others to live lives of infamy. Violence took most of these lives, as many as two per day, not even removing their boots – hence the name, "Boot Hill."

Boot Hill

Wm. Coffman, shot 1875

$75.00 Photo

WE WERE HEADED FOR A 9:30 A.M. appointment in Shelton, Nebraska. It was a bright Nebraska morning, with the rising sun in a clear sky—photogenic perfect. Just ahead were the Union Pacific tracks and to the right an ideal place to park. Surely a train would be along soon to pose for a portrait. We didn't put our ears to the tracks, but we did peel our eyes. After ten minutes we decided our appointment was the better part of valor, crossed the tracks, and turned left on the Lincoln Highway for Shelton.

A mile and a half later, glowing yellow in the morning sun, was a Union Pacific diesel headed our way followed by cars as far as you could see. A quick U turn and we were headed back hoping to beat the train for our photo op. As we crossed the tracks to park, blue lights flashed in the rear mirror. I explained to the officer I was doing some Lincoln Highway research and the light was just right for a photo. "I clocked you at 71 miles per hour," he said, and then asked the usual: driver's license, owner's registration, etc. The registration was in a cubbyhole in back mixed with some trailer papers. By now the train was closing in.

"Give me the papers," said the officer. "You go get your picture and I'll find what I need." The train came and went and I got my picture – along with a complimentary speeding ticket. "We cooperate with other states," he said, "so your $75 fine will be waived if you take a defensive driving course when you get home." The necessary phone calls and arrangements were made and my costs tallied up to $50 court costs and $25 for the driving class = $75.00

OMAHA, *Nebraska*

Russell Rein

IN HIS OWN WORDS:

Sunday, September 5, 1915

    Arrived in Omaha, Neb. 56 miles good dirt roads Bad hill 6 miles west of Council Bluffs.

Monday, September 6, 1915

    Rain. Labor Day. Hunting place to set off Big body until return from coast. Hikers left us.

Tuesday, September 7, 1915

    Fair. Ging arrived. Hunted camping place and also place to set off truck.
    Set off body at 1835 N. 20th St. Omaha, Neb.

Wednesday, September 8, 1915

    Fair. Put off body. Had hard time of it.
    Have ordered Prairie Schooner body at Johnson Danforth Co. N. 16, city.

Thursday, September 9, 1915

    Fair. Working on new top.

Friday, September 10, 1915

    Fair. In Omaha all day.

Saturday, September 11, 1915

    Fair. Got Prairie Schooner Top. Left Ike [an employee] in Omaha.

[November and December visits to Omaha were on his way back East]

Saturday, November, 13, 1915

    Left ? with truck and got $11.00 to get into Omaha, Neb.

Sunday, November 14, 1915

    Arrived in Omaha Neb. 6:00 A.M. Located at 1524 N. 18th st . until truck is fixed.

Monday, November 15, 1915

    Fair. Got out and made 12 shots in city. Come very hard.

Tuesday, November 16, 1915

    Cold. Passed and delivered what was shot Monday. Very bad average.

Wednesday, November 17, 1915

    Cold. Made a few shots in Omaha.

Thursday, November 18, 1915

    Passed proofs and made delivery of couple dollars.

Friday, November, 19, 1915

    Cold. Nothing doing at all. Got ready for Sat. shooting.

Saturday, November 20, 1915

    Fair. Went out and shot up Florence Made 14 shots on kids and jobs.

Sunday, November 21, 1915

    Cold. In room all day. Run off proofs and made trip photos.

Monday, November 22, 1915

    Fair. Run down to Papillion, Neb. to work. Town was shot last week. Packed up things for trip Tuesday

Tuesday, November 23, 1915

    Left city 7:30 A.M. for Missouri Valley, Logan, and Woodbine, Iowa. All three had just been shot lately and nothing doing west (back) to city.

Friday, November, 26, 1915

    Arrived in Omaha at 2:00 P.M. 80 miles.

Saturday, November 27, 1915

    Cold. In city all day. Set body off at rear of Oheary's yard on 20th St. and took chassis to N. 16th to make repairs on spring.

Sunday, November 28, 1915

    Cold. In city all day. Sold pyrene to get money for repairs on spring.

Monday, November 29, 1915

    Cold. Went out and tried to shoot. Got 7 shots and got car out of shop.

Tuesday, November 30, 1915

    Cold. Passed proofs. 7 that were made Monday. Passed very bum.

Wednesday, December 1, 1915

    Cold. Delivered and made 2 shots in ? day. Very rotten.

THE LAND WHICH BECAME THE CITY of Omaha was part of the Louisiana Purchase, completed by Thomas Jefferson in 1803. The word "Omaha" in native language means "above all others upon a stream" or "against the current." It is thought that the Omaha Indians got that name because of their northward travel against the current of the Missouri River.

The city was first laid out in 1854 by the Council Bluffs and Nebraska Ferry Company. In the beginning lots were given away to whomever would make improvements on them, but by June 1855, the population had grown to 250, with lots selling for $100 each. By 1857, two years later, the population was 1,500 with lots selling for $4,000. The first resident of Omaha, William P. Snowden, built a crude log building which eventually became the St. Nicholas Hotel. Omaha was officially incorporated by the state legislature on February 2, 1857.

The experience of most pioneer endeavors was not without its bitter-sweet, with Omaha being no exception. Here it was the Mormons who established the first non-native settlement in Florence, just north of Omaha, with 1,000 houses as part of their movement west to Salt Lake City. During their two-year stay, 1846 to 1848 they suffered many deaths due to inclement weather, poor living conditions and lack of proper food. A cemetery and museum bear witness to their pain.

The strategic role of railroads is evident again in Omaha. December 2, 1863, is a date worthy of note. The events of that day accelerated the growth of Omaha and illustrated again the influential presence of Abraham Lincoln. That was the day ground was broken for the eastern terminus of the new transcontinental railroad, just

1835 N. 20th Street

1835 North twentieth Street is where my father set off the body of his truck to be replaced by a prairie schooner body for the trip west. Obviously, 1835 has changed considerably during the past eighty-five years!

fifteen days after President Lincoln declared the eastern terminus to be Omaha on November 17. A celebration at the Herndon House followed the groundbreaking. The railroad was now poised to link the Atlantic and Pacific coasts for the first time. Fifty years later, in 1913, the Lincoln Highway provided the second continental link, unleashing forces which eventually eclipsed even the greatest railroad.

In 1915 my father saw a lot of Omaha—on three separate occasions in fact. He first entered town on Sunday, September 5, 1915. He arrived with several Philadelphia hitchhikers from Dunlap, Iowa. He reported, "Good dirt roads" but a bad hill six miles west of Council Bluffs, probably Honey Creek. On Monday, September 6, he looked for a place to set off the body of his truck to install a covered schooner body, presumably for travel across the desert. The body was set off at 1835 N. 20th Street and the prairie schooner body installed at the Johnson Danforth Company. This process took the whole week. He left Omaha September 11 for Valley, Nebraska.

On his return from San Fransisco in November he had the misfortune of a broken clutch in Lexington, Nebraska, and the truck was towed to a garage in Overton. He then took the train to Omaha where he stayed at 1524 N. 18th Street until the truck was fixed. He doesn't say what he used for transportation, but he did travel to Omaha, Florence, Papillion, Missouri Valley, Logan, and Woodbine where he took some photos. He records the weather as "very cold." He left Omaha by train November 24 for Overton to pick up his truck.

We see him the third time in Omaha on Friday, November 26, arriving from Schuyler. The next day he set off the prairie schooner body at the rear of Oheary's yard on 20th Street and took the chassis to Johnson Danforth on North 16th street for spring repair. He stayed in town until Wednesday December 1, plying his trade with very poor results. His words: "very rotten."

The last we see him in Omaha is Saturday night, December 4, coming by train from Dunlap. His photography must not have been very successful because on Sunday he wired home to Lancaster County, Pennsylvania for money to pay for truck repairs. On Monday morning he left Omaha for the last time and headed back to, Dunlap, Iowa.

VALLEY, *Nebraska*

IN HIS OWN WORDS:

Saturday, September, 11, 1915

Fair. Booked Valley for Tuesday. 93 miles from Omaha."

Sunday, September, 12, 1915

Fair. Camped at Bath Lake, Valley, Nebraska.

Monday, September, 13, 1915

Fair. Shot up Valley for show Tues., and Wed.

Tuesday, September, 14, 1915

Fair. Run slides for Valley and proofs. Show went $13.50, 50% 50%

Wednesday, September, 15, 1915

Fair. Passed proofs and delivered in Valley. Sold slides to manager for $2.00.

MY FATHER INTRODUCED HIMSELF to Valley with his camera on September 11, 1915. In fact, he stayed in Valley five days photographing the children of Valley residents. He even booked the local theater for Tuesday September 14 and Wednesday September 15, to show off these children on the screen.

The weather was "Fair" with no rain and no floods during those five days. This was not always the way it was in Valley. High water was a wearisome topic for residents since the town lay between two rivers, the Platte River on the west and Elkhorn on the east, making Valley no stranger to floods. In pre-dike days, it was not unusual for spring rains to swell the Platte and Elkhorn over their banks, covering all the land between them. One positive economic consequence was the abundant supply of sand and gravel, a byproduct of the last glacial age. The sand pits provided not only a generous supply of quality building material but also beautiful lakes for recreation. My father camped at Bath Lake.

Valley Community Historical Museum

March, 1912 Flood

West Whittingham Street. Water was reported to have been three to four feet deep in these houses.

Valley Community Historical Museum

Opera House

Valley Theater (Gem Theater) was the site of the Opera House built in 1892 by a stock company selling shares at $100 per share. $2,000 was raised for its construction. No doubt this is the theater booked by my father for September 14 and 15, 1915. The theater burned to the ground October 31, 1947.

Bank of Valley Corner

Called Farmer's State Bank, it was built in 1914, a new structure when my father visited Valley in 1915.

Valley Community Historical Museum

The first flood of note occurred in 1872. A Mr. Talcott had gone to Omaha on business and on his return was cut off by rising waters of the Platte River. He was forced to make a circuit to the north, finally arriving home three days later by boat.

Another flood recorded in 1881 resulted in seventy-five people seeking refuge in a grain elevator. It was during this flood that residents witnessed a small church float by from somewhere in South Dakota with its bell tolling as it passed.

On the morning of March 29, 1912, residents of Valley looked out their windows upon an ocean of water reaching all the way from Elkhorn to the Platte River. This flood was made even worse by an ice gorge northwest of town which diverted the flood waters toward the Union Pacific Railroad tracks. The water followed a course downtown destroying many miles of track. In 1919 a dike was built along the Platte River to avert such catastrophes.

In 1864, Valley was laid out on part of a federal land grant to the Union Pacific Railroad in the Platte River precinct. The first settler, John Saunders, and the first town resident, Richard Selsor, both arrived that same year. In 1875 the town boasted a hotel, a general merchandise store, a school and a Methodist church. By 1890 a newspaper and bank were added to its growing list of successful enterprises.

In 1913 Valley became one of the forty-seven hosts of the Lincoln Highway in Nebraska. In 1925 it inherited the numerical designation of Route 30. Today Valley is a delightful Lincoln Highway byway in the shadow of Omaha and Elkhorn.

SCHUYLER, *Nebraska*

IN HIS OWN WORDS:

Thursday, September 16, 1915

Arrived in Schuyler, Neb. 8:00 A.M., 15 miles. Booked for Friday 17th and shot the town. 60 shots.

Friday, September 17, 1915

Rain. Run off proofs and slides for Schuyler. Had bad rainy day for show. Rec'd $14.60 50% 50%. $7.30.

Saturday, September, 18, 1915

Fair. Passed proofs - $12.00. Store fronts will show slides again Monday night.

Sunday, September 19, 1915

Fair. Run off orders and made shot at garage. Camped in town all day.

Monday, September 20, 1915

Fair. Run slides for show this night. Receipts $27.60 50% 50%. Camped west of town.

Tuesday, September 21, 1915

Fair. Frost. Left west Schuyler 7:30 a.m.

[on his way back East]

Thursday, November, 25, 1915

Traveled to Schuyler, Neb. Stayed all night at Farmer's Hotel. Made 100 miles this day.

Friday, November 26, 1915

Fair. Left Schuyler 7:00 A.M.

SCHUYLER WAS NAMED FOR SCHUYLER Colfax, vice-president of the United States during the administration of President Ulysses S. Grant. In 1847, by accident, fate or timing, Schuyler entered a page of history by being located along the Mormon Trail where traveled thousands of Mormons on their way to Utah.

There are two things that usually rise to the historical surface of Lincoln Highway towns: rivers and railroads; the one by providence, the other by entrepreneurship. In this case "providence" was the Platte River, the crossing of which was first accomplished by using Moses Shinn's ferry near the mouth of Shell Creek, and later in 1866, "entrepreneurship" was the Union Pacific Railroad which established a depot called "Shell Creek Station."

It's a wonderment how strange events often affect the future. Buchanan, an early settlement just north of the Platte River, was bypassed by the railroad. Mail delivery therefore was infrequent and usually at night. The mailbag for Buchanan was tossed out in the general

Vine Street

Vine Street (1915), now 11th Street. The far left building is the former Kopac Brothers Garage, now Department of Utilities. On the right is the former City Hall, now the police station.

City Hall, Schuyler

City Hall on the Lincoln Highway, is now the police station. President Ben Vrana provided this old photograph from the Schuyler Museum archives.

area and it was the postmaster's responsibility to search for it the next day. A petition was circulated to close the Buchanan post office after a band of Indians came up the tracks trying to sell the contents of a mailbag to section workers. Dan Harshberger, from the Shell Creek Station, took the two-by-three-foot post office box. (That's right, Two-by-three-feet! Now we know why the Union Pacific bypassed Buchanan.) Let's assume that Dan Harshberger gave it a good home in Shell Creek and used it as seed for the later Schuyler post office.

In 1869 Platte County was divided into three smaller counties. Colfax County was the easternmost section, with Shell Creek as county seat. It was renamed Schuyler and consisted of only a section house, a water tower, and a place to store fuel until the Smith brothers opened a small general store.

In 1870 Schuyler began to flex its muscles and was incorporated. It was no accident that in the same year Texas cattlemen began driving large herds of cattle to the Schuyler stockyards. This caused the town to grow to more than 600 residents and more than 100 businesses, including livery stables, a brewery, and other enterprises to serve the cowboys. During 1870 alone over 40,000 head of cattle went through the Schuyler stockyards.

This prosperity, and seemingly endless booming economy, came to an unfortunate end in a few years, (with no fault of the people of Schuyler) when a severe drought and an invasion of grasshoppers brought the economy to its knees. If that were not enough, during a severe late-night thunderstorm several hundred cattle were spooked, and hungry settlers south of the river killed and butchered as many of the estranged cattle as they could find for their own consumption. Schuyler suffered even more when the Texas cattlemen began driving their cattle to stockyards farther east.

The rains of 1877 began to turn things around. From 1870 to 1890 there was an influx of Czech, Irish, Scots, and German settlers, to whom the railroad sold land for three to eight dollars per acre. No dummies, they. They, too, had heard the siren call of the West and were not shy about snapping up a good deal when they saw it. Schuyler, on the main line of the Union Pacific Railroad, continued to experience steady growth ,so that by 1900 it

Ford Garage

1915 Lincoln Highway at the intersection of 11th and A Streets.

120

IOOF Building

The Lincoln Highway in 1915 jogged at this intersection of 11th (Vine in 1915) and B Streets. The structure is the IOOF building.

could boast of a population of 2,127 and 100 businesses. It was into this prosperity that my father entered September 16, 1915. He booked a local theater for September 17 and again for September 20, spending six days in town, leaving the 21st at 7:30 A.M. for Grand Island. On November 25 on his way back east, he again stopped in Schuyler and stayed over night at the Farmer's Hotel.

The city of Schuyler took seriously the coming of the Lincoln Highway, evidenced by the following item in the June 1, 1914 issue of;

THE SCHUYLER SUN

The two big Lincoln Highway signs ordered some time ago by the local association have been made, and the first of them was placed in position in the west edge of town Wednesday. The other will be put up later this week, since it had not been completed at the time the other was ready. The signs read "This is Schuyler, Nebraska." And the distances to New York, San Francisco, North Platte and Omaha are also given. Another feature of the signs is the line, just beneath the name of the town, "population 3,000 and still growing."

The signs are ten feet by five, and are painted in the highway colors of red, white and blue. The lettering on the white field is black. They are made from wood, covered with heavy tin, and are coated with several coats of paint and varnish, which will make them very durable and weatherproof. Electric signs will light them at night, turned off and on from the powerhouse with the street lights. They are placed in such positions that every corner can see them plainly, and will be "good business" for the town.

The Schuyler population in 2000 was 5,371.

SHELTON, *Nebraska*

IN HIS OWN WORDS:

Wednesday, September 22, 1915

Fair. Arrived at Shelton 10:00 A.M. Booked for Thursday night.

Thursday, September 23, 1915

Fair. Showed stores and schools. Had $21.20| House. 50% 50%. Delivery $4.75.

Friday, September, 24, 1915

Fair. Delivered. Left 10:30 A.M.

SHELTON IS LOCATED ALONG THE Wood River in the extreme eastern part of Buffalo County. *Andreas' History of the State of Nebraska* identifies it as, "A pleasant little town located on the right bank of the Wood River." In 1858 it was the center of the Mormon settlement, and is also where B. F. Sammons started his Mormon newspaper, *The Huntsman Echo.* Later it was the residence of Patrick Walsh, who became the most colorful politician of the era in Shelton. The original post office was called Wood River Center, but just a few miles east in Hall County was another Wood River. This duplication of "Wood Rivers" in close proximity resulted in confusion with mail delivery and set the stage for the name change to Shelton.

Patrick Walsh, self-appointed postmaster, was so upset with two Wood Rivers he took it upon himself to send a cryptic note to the Postmaster General, "To the Postmaster General: Sir—you are hereby notified that as there is another Post Office in the state named Wood River, the name of Wood River Center is changed to Shelton, and you will please govern yourself accordingly," and signed

Meisner Bank, Shelton, Nebraska

On the corner of D Street and the Lincoln Highway. The ground floor is now the Shelton Historical Interpretative Center, and the theater where my father showed his pictures of local children and businesses was to the right of the bank. Note the Lincoln Highway sign on the lamp post.

Jane Bernhard stands among her flowers in front of her Shelton home, on her eighty-third birthday. She is the great-granddaughter of Nellie Meisner.

his name officially as postmaster. The Post Office Department changed the name without further question. Shelton has been Shelton ever since.

In 2001 when Miriam and I visited Shelton we were privileged to meet Jane Bernhard, a descendant of the Shelton Smith family. She told us the story of Shelton's Meisner Bank. George Meisner was born in Germany March 19, 1843 and came to this country with his parents in 1847. He came to Shelton via Troy, New York, and Tama, Iowa, arriving in Shelton in 1871 at age 28.

He homesteaded eighty acres the first year, 240 acres two years later and eventually farming 3,000 acres of corn, wheat, cattle and sheep. He married Rachel Fieldgrove in October, 1877 and they had four daughters, Eldora, Elnora, Cora and Lulu.

In 1882 the Shelton Bank had opened with J. M. Coleman as president. The bank was an important institution for the Shelton community, made prosperous by fertile soil and industrious farmers like George Meisner. As the bank continued to prosper, fueled by easy access to markets via the Union Pacific Railroad, George Meisner became available and willing to take control.

In 1909, when Meisner was sixty-six years old (that's important) he built the Meisner bank building, a large imposing building on the corner of D Street and what became the Lincoln Highway four years later. Meisner, understandably, took a keen interest in the bank's construction, often offering not always appreciated suggestions. Near the completion of the building he became so upset, with his perception of slow progress, that he volunteered to show the workmen how it should be done and even carried brick up the ladder himself. That night he had a heart attack and died a few days later. Thus it is today that one can read the coping at the top of the bank building, "Meisner—1909," the year of George Meisner's death. The ground floor of this building now houses the Shelton Historical Interpretative Center. In 1915 the theater where my father showed photos of local children and businesses was next door to the bank on D Street.

After George Meisner's first wife died, he married twenty-eight year old Nellie who was only ten years older than George's oldest daughter, Elnora. After George's death, Nellie traveled and spent lavishly. She would go to Omaha and New York to outfit herself with the finest and latest. It was said that the train from Grand Island to

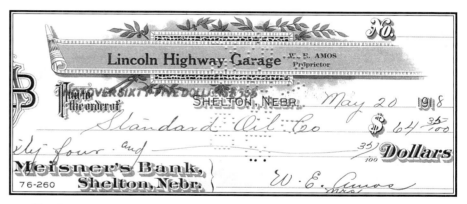

Check drawn on the Meisner Bank by the Lincoln Highway Garage, May 20, 1918.

Kearney would slow down through Shelton so everyone could see the latest styles. "Nellie was spending money as though she had an oil well that would never go dry" and "Nellie didn't even know how to spell 'money'," were good-natured local assessments of George Meisner's second wife.

Shelton is still, "A pleasant little town located on the right bank of the Wood River"—along the Lincoln Highway.

Jane Bernhard

Muriel Bernhard

Muriel Bernhard, Jane Bernhard's mother-in-law, stands by the family car with her two daughters, Delores and Marcella.

Jane Bernhard

Fixing a Tire

Anna and Pete Petersen invited Bess Smith and Evan Bernhard, Jane Bernhard's parents-to-be, to join them on a trip to California in 1909. Aunt Nellie Meisner loaned them her new Rambler for the trip. Anna and Pete said to Bess and Evan, "Why don't you two get married so we can have more fun?" So they did, in Fort Collins, Colorado, while on the trip.

Jane Bernhard

Nebraska Sheep

Marcella Bernhard stands among Shelton, Nebraska, sheep in 1916.

KEARNEY, *Nebraska*

IN HIS OWN WORDS:

Friday, September 24, 1915

Arrived in Kearney 2:00 P.M. 18 miles.

Saturday, September, 25, 1915

Rain. In all day in Kearney waiting for victory films to be shipped from I. Morrow, Omaha. Received films 5:30 P.M. Left town. Camped 20 miles out.

Sunday, September 26, 1915

Rain. Left camp 9 A.M.

OFTEN IN HISTORY CERTAIN PERSONS rise head and shoulders above the rest. For Americans it was George Washington, and later Abraham Lincoln. For Kearney, it was the Rev. A. Collins. He was a prominent presence in most every important early event of Kearney: his was the first family, May 11, 1871; first postmaster, 1871; officiated at the first death, 1872; minister of the first church, October 1871; Rev. Collins and his wife, Louisa, were among the first five members of the church; the first Sunday school was in the home of Rev. Collins, February 25, 1872; the first lot of the first platted land by the Burlington & Missouri Railroad was presented to Mrs. Collins in September, 1872. He performed the first marriage, 1872; Rev. Collin's son the victim of the first murder, September 17, 1875; the first Judge.

For Kearney, the major events which make a community, were compressed into just two years—1871 and 1872.

D. N. Smith, who was in charge of the anticipated town site for the Burlington & Missouri Railroad, arrived there in a three-day blinding snowstorm in early April 1871. A survey was made followed by the decision to lay out a town site at this location between Elm Creek and Kearney Station.

The tall man of Kearney was the Rev. A. Collins.

The tall years of Kearney were 1871-1873.

The dark years of Kearney were 1874-1876

The Nebraska Public Power District condenses the history of Kearney very well.

The city of Kearney derived its name from the original fort but due to a postal error an "e" was inadvertently added and then never changed.

Kearney Junction began a period of rapid growth increasing from 254 residents in 1873 to well over 10,000 in the late 1880s. Optimistic residents sought to have the nation's capital moved to Kearney from Washington, D. C., and others raised a quarter million dollars to finance the construction of a huge cotton mill.

The bubble burst in the 1890s. The cotton mill was closed, real estate values collapsed, businesses and people drifted away. In 1900, only 5,364 people remained.

In the early twentieth century, Kearney began a steady, if not dramatic, recovery. By 1930 the population had increased to over 8,500 and the community was laying the foundation of its present diversified economy.

The University of Nebraska at Kearney, originally Kearney Normal School, was founded in 1905. Today, with an enrollment of nearly 7,000 students it is the state's premier residential undergraduate teaching institution.

Kearney, population 27,000+, now boasts a balanced thriving economy and is among the fastest growing cities in the state.

The darkest days of Kearney began with the conflict between farmers and cattlemen. It could be argued that cattlemen had the advantage of prior occupancy with their large herds of stock grazing at will over many miles of prairie grass, long before the arrival of farmers. Then came the farmers with plows and fences incompatible with grazing horses and cattle.

Lincoln Highway Station, Grand Island, NE

733 Ranch Sign

A circa 1921 Ford stands in front of the 1733 Ranch sign.

The issue climaxed in 1874 as conflicts increased. Horses and cattle damaged the farmers' corn and other crops and the cattlemen refused to pay damages. Rowdy herders began showing up in Kearney with specific intent to bully and terrify settlers. The issue came to a head on September 17, 1875, when some ponies belonging to Jordan P. Smith created considerable damage in the cornfield of Milton M. Collins, son of Rev. Collins, who put them in a corral until the owners would come for them. This tragic story ended when Jordan Smith, with three more cowboys, rode up drunk and commanded Collins to turn the ponies out. When Collins began to get off his horse to release the horses, Smith shot him in the heart and several more times as he lay on the ground.

Smith was captured shortly thereafter on an island of the Platte not far from Plum Creek. He was tried in Kearney, found guilty and sentenced to death by hanging. However, due to some legal technicality, Smith received a change of venue to Lowell, Nebraska, and again to Juniata, where he was found guilty of manslaughter and sentenced to ten years.

As is usually the case, time finds a way to triumph and so it was in Kearney. The trauma of these events gradually cleared peoples' heads, with law and order triumphing over the dare-devil, half-drunken character of the cowboys.

The above has little to do with the Lincoln Highway except to pull aside the curtain of time to give a peek at the people who gave birth to this city on the Lincoln Highway.

My father arrived in Kearney on Friday, September 24, 1915, in the rain, and left three days later in the rain, on September 26, with no comment about muddy roads. One would think that rain, roads, and Nebraska are not a good recipe for charming auto travel. However, in the case of Kearney, with its close proximity to the Platte River (which provided a light sandy soil), rain did not always adversely affect road surfaces.

In the context of the Lincoln Highway, roads and cars, Kearney made a stab at history on another front—auto manufacture. Kearney was host (well, almost) to the manufacture of two autos:

Deserted Gift Shop

Deserted covered wagon gift shop, 2.5 miles west of Kearney on the Lincoln Highway. The two elderly proprietors had moved to a nursing home a few days before we stopped.

Bolte, 1900 by Thomas H. Bolte, owner of a bicycle shop. He intended to build the car in Denver, but his company turned to cement mixers instead.

Kearney, 1907 by the Kearney Foundry, Machine & Automobile Company. It is doubtful that any cars were ever built.

Cottonwood Grove

Cottonwood grove 3.5 miles west of Kearney on the Lincoln Highway. The cottonwood was a staple tree for early pioneers.

The Arch, Kearney, Nebraska

Great Platte River Road Archway Monument arches over Interstate 80. "Whatever has been done by other displays on other monuments to depict the Westward flow in the era of Manifest Destiny, the Nebraska monument does it better." Bill Leonard, Des Moines Register.

Kearney, Nebraska

In front of the Nebraska Department of Roads, Kearney, Nebnraska.

Kearney, Nebraska

Lincoln Highway Post at the Great Platte River Road Archway Monument, dedicated June 14, 2001

ELM CREEK, *Nebraska*

IN HIS OWN WORDS

Sunday, September 26, 1915

Arrived in Elm Creek, Neb. 17 miles. Bad muddy roads. Had H. of a time getting through. Camped in town.

Monday, September 27, 1915

Fair. Putting into town this A.M. broke part of clutch. Was towed to garage. Wired for part to Chicago. Showed Victory. Recd. $21.15.

Tuesday, September 28, 1915

Fair. In Elm Creek all day. Shot town got 12 shots on stores.

Wednesday, September 29, 1915

Fair. Made up proofs and canvassed them for delivery. Truck part not received yet

Thursday, September 30, 1915

Delivery few photos, 1 day, .75.

Friday, October 1, 1915

Fair. Received clutch part. Got fixed up and left at 3 P.M. Traveled beyond Lexington 29 miles. Rough roads.

R AILROADS, WATER AND WOOD WERE the magnets which drew a steady stream of Irish immigrants, among others, to Elm Creek. The town was born in August 1866, when the Union Pacific Railroad laid its tracks through Buffalo County and built a siding along a creek bordered with elm trees, which inspired railroad workers to call the creek "Elm" and the siding, "Elm Creek Siding." Later when a station was built, it was called "Elmcreek" station, one word. Later it was officially divided into two words.

Elm Creek was incorporated January 12, 1887, with a population of 300. The thriving community was serviced by three hotels, a restaurant, bank, flour mill, drug store, five general stores, two livery stables, two farm equipment stores, two hardware stores, three meat markets, two lumber companies, a post office, two lawyers, one doctor, two churches, a school and a newspaper.

Tragedy struck July 1, 1906 when a fire destroyed fourteen twenty-year-old frame buildings on Front Street. Replacement buildings were of stone with brick fronts. Like most towns in the Platte River Valley, Elm Creek

Helen Casper

Elm Creek, Nebraska

Elm Creek, as my father saw it in 1915. Note the garage on the corner, left center. It looks the same today.

endured its share of floods, mostly south of the tracks. Exceptions were in 1932 and again in 1947, when both Turkey Creek and Elm Creek flooded even north of the tracks.

Modern Elm Creek boasts a population of over 900 with the intersection of Highway 183 and the Lincoln Highway (Route 30) to provide access to Elm Creek markets for passers-by and residents to nearby trading areas.

Grinding through Nebraska mud took its toll on my father's Little Giant truck. On Sunday, September 26, 1915, he wrote, "Arrived in Elm Creek, Neb. Bad muddy roads. Had H. of a time getting through." The next day

Ford Garage

Ford garage owned by the late Paul Casper on the Lincoln Highway in Elm Creek. It is the same garage where my father had the clutch fixed in 1915.

Theater

In 1915 this was the theater where my father showed local pictures on Monday, September 27, 1915.

part of the clutch broke and he was towed to a garage in Elm Creek where he wired Chicago for the part. While the truck was laid up, he photographed children and businesses, and rented the local theater to "show them off." On Friday, October 1, 1915, the clutch part arrived, the truck fixed and he departed yet that day at 3:00 P.M. This event was in September on his way west, and on November 12, on his way back, the clutch broke again—this time also in Nebraska, at Lexington near Elm Creek. The truck was towed to a garage in Overton.

OVERTON, Nebraska

IN HIS OWN WORDS:

Friday, November 12, 1915

Towed machine to Overton, Neb., And got money for train fare

Wednesday, November 24, 1915

Fair. Left city [Omaha] 12:40 A.M. for Overton, Neb. for truck. Helped make repairs and left 5:00 P.M. Traveled until 1:30 A.M. Thursday to Shelton, Neb. Stopped at hotel.

Thursday, November 25, 1915

Cold. Left Shelton 7:30 A.M.

MY FATHER'S EXPERIENCES IN OVERTON were shared with Lexington, Shelton, and Omaha. On November 12, 1915, with the aggravation of Nebraska mud, the clutch on his machine broke fifty miles east of Lexington and the truck was towed to Overton. He then took the train to Omaha where he plied his photo trade for the next 11 days while waiting for the part to come from Chicago. When it came he returned to Overton, helped fix the truck and took off at 5:30 P.M. for Shelton, arriving at 1:30 A.M. He checked into a hotel.

In the 1860s the Union Pacific Railroad established a siding in Dawson County named Overton in honor of either a railroad dignitary, section foreman, or government official guarding railroad workmen. The first depot was a two-section box car, with one end a ticket office, the other, living quarters. At that time Overton was little more than a station house, post office, a couple houses, and a shiny set of rails with scattered farmsteads in the Platte Valley. The railroad employed many farmers who supplemented their early meager returns from breaking sod. Farming, however, eventually moved Overton

Mildred Heath,
Editor, Overton Observer

Heath, editor for seventy-five years, stands in the doorway of her office. We followed her down an alley, to the rear entrance of a senior citizen facility, for lunch.

beyond a railroad town, since it was the major vocation for immigrants from Ireland, English-Americans, and Civil War veterans. Among this mix were blacks, who traveled the Underground Railroad to Canada, then returned to the United States after the war, some of them settling in Overton. Many of these refugees were black women who married white men in Canada. One of these was Charles Meehan, of Irish descent, who married a black woman in Canada and spearheaded the arrival of about twelve black families to Overton, attracted by homestead laws. Mildred Heath, editor of the Overton Observer, describes them as "…good neighbors [who] strove to make their life and community a better place to live. Deeply religious, they worked hard in the church, and saw that their children were well educated, many going on to college."

Overton became a village in 1873. In 1874 a Mr. Boles built a sod-frame building which also served as post office and grocery store. Also in 1874, a school was built which also served as a church and community building. Overton still sits astride the Union Pacific Railroad, serving a thriving alfalfa and grain community while at the same time hosting the growing number of Lincoln Highway fans who are convinced it's time well spent to exit I-80 (only two miles south) to experience the amiable hospitality of the people of Overton.

Overton Observer

Buick

1911 or 1912 Buick on the street in Overton. Note one of Overton's grain elevators to the right.

Overton bridge

Overton's Lincoln Highway bridge with Union Pacific coal cars passing nearby.

Overton Observer

Garage

My father did not say, but his Little Giant truck could have been towed from Lexington to this garage in Overton for clutch repair.

Nebraska State Historical Society

Overton Herald

Overton Herald in 1904. Was this the forerunner of Mildred Heath's Overton Observer?

PAXTON, *Nebraska*

IN HIS OWN WORDS:

Saturday, October 2, 1915

Fair. Left camp at 6:30 A.M.. Traveled to Paxton, 95 miles, good roads.

Sunday, October 3, 1915

Cold north west wind. Broke camp at 1:00 A.M.

WHAT'S IN A NAME? QUITE A BIT if you want potential settlers to take your "Y'all come" seriously and the name of your town is Alkali. At least that was the logic of the town fathers of Alkali, Nebraska, in 1885 when they changed the name from "Alkali" to "Paxton". Prior to the 80s the surrounding grasslands were grazed by large herds of cattle, one such owned by William Paxton, an Omaha cattleman with numerous other local business interests. His name, Paxton, received the selection committee's final endorsement—henceforth this community would heed the call of "Paxton." Alkali is a mixture of soluble salts which when in quantity are detrimental to agriculture, especially in dry weather. Alkali in Paxton was so severe that huge water softening equipment operated twenty-four hours a day, filtering the water through excelsior, lime, and soda ash to prevent foam in the steam engines of the locomotives.

Many private dramas unfolded as homesteaders moved into the area. In 1884 a widow, Mrs. Ann LeDioyt, from Wilmington, Illinois, filed one of the earliest claims in the area. She and her five children arrived with all their belongings in a covered wagon and filed a quarter-section claim (410 acres) adjacent to a quarter section belonging to a relative. The two families had a problem. What could they do about the claim rule, which required each claim to have a building on it? Widow LeDioyt and her relative exercised some "Yankee" ingenuity by building a single room frame house straddling their two quarter-sections and hanging a curtain down the middle of the room, dividing it for the two families while at the same time satisfying the claim rule.

On our visit to Paxton in 2001, my wife and I did not meet any of the widow LeDioyt descendants. We did, however, learn to know Fae Christensen, a descendant of early Paxton settlers from Sweden. She is the unofficial town historian and has many boxes and files of records in her home to prove it. Her husband, Orville, was Mayor of Paxton in the 1980s. We found Fae a gracious and enthusiastic fan of Paxton and promoter of its heritage. Concerning the early years of Paxton she said, "These were incredible years. History records that in 1872 a herd of buffalo in North Platte valley stretched from O'Fallon to Ogallala (thirty-two miles); 200 imigrant wagons passed by on the Oregon Trail during the week of June 5,

Fae Christensen

Fae Christensen stands with Gene Jean at his home in Paxton on the Lincoln Highway.

Paxton, Nebraska

180 degree view of Paxton to the east and the west, 1920.

1875; and seven rail-carloads of imigrants per day passed through during the week of October 9, 1875. In August 1876 trains were delayed for several hours by grasshoppers that darkened the sky and covered the ground, tracks and all."

A decade of much rain made farming attractive during the '80s and Paxton grew rapidly. This expansion came to a halt, and actually declined during the drought of the '90s. Many people gave up and moved away. The Bank of Paxton opened in 1888 and closed in 1896. By 1900, the rains returned, along with a return to prosperity. Paxton's prosperity again declined during the drought, depression, dust and destructive grasshoppers of the 1930s, from which the town has never fully recovered.

Fae summarizes her feelings about her Paxton: "Today, as the wagon ruts of the Oregon Trail disappear in the sod, they are replaced by contrails of the jets as we look heavenward and to the future, knowing full well that 100 years hence, Paxton will still be here."

Paxton Garage

The Dafler building was built in 1919 – auto repair upstairs. Note the Lincoln Highway marker in the street.

Ford Garage

Former Paxton Ford garage as it appears today.

"Off" The Lincoln Highway

Fae's father, Carl Traulsen (right), needs some assistance to get his new Buick back on the road.

Courtesy, Fae Christensen

Bridge

An early bridge across the South Platte River at Ogallala, 1886-1887.

Fae Christensen

Nissley Family Collection

Alkila Flat

"Alkila" is also "Alkali" if the "a" and "i" are switched. This is most likely what happened when my father marked the negative of the above photo. Since the first name for Paxton was Alkali, Paxton is likely where the photo was taken. He stands on the running board of his Little Giant, having just recently had an encounter with Nebraska mud.

Fae Christensen

Mutual Aid On The Lincoln Highway
The feet of Carl Traulsen can be seen extending from beneath a disabled car, 1919. The Traulsen family, with neighbors Hege and Schomer, are on a three-day trip to Omaha. They traveled together for mutual assistance. Fae Traulsen (Christensen) is the little girl in the center. Her mother, Eva (with big hat), looks down at her.

Spruce and Third
Northwest corner of Spruce and Third, Paxton, Nebraska

Oak And Third
Southwest corner of Oak and Third, Paxton, Nebraska

SIDNEY, *Nebraska*

IN HIS OWN WORDS:

Sunday, October 3, 1915

*Cold NW wind. Traveled to Sidney,
105 miles through sand and prairie roads*

Monday, October 4, 1915

*Cold, fair. Run in town. Had a hard time getting
out of sand where we camped. Booked Victory
at Sidney and showed. Received $31.60, 40% $12.60.
Camped out of town.*

CHICAGO HAD ITS AL CAPONE, New York its Mafia, but Sidney had Calamity Jane, Wild Bill Hickcock, Sam Bass, Whispering Smith, Doc Middleton, Butch Cassidy, Canada Bill Gang, Dr. Baggs, Jim Bush, Jim Lavine, and Rebel George. Sidney was a lawless place, wide open with gambling, murder, and general hell-raising, spawned by daily shipments of gold from the Dakota Territory. There were twenty-three saloons in one block, with eighty-nine establishments selling liquor. The effect of this absence of *corpus juris* was felt beyond the city limits of Sidney, and in 1882 the Union Pacific Railroad took a hand to help rid the community of its outlaws.

Sidney began in 1867 with the laying of tracks for the Union Pacific Railroad across western Nebraska, inciting opposition from the Sioux, Cheyenne, and Arapahoe. In order to protect workers from attack, military units from Fort Sedgwick and Cheyenne provided escorts, station guards, and scouts. In 1867 a troop station was established at Sidney and thus began the twenty-six year history of Fort Sidney, which ultimately grew into a complex of some forty buildings. The coming of irrigation increased the flow of homesteaders and decreased the need for military garrisons. Fort Sidney was abandoned on June 1, 1894, and the buildings sold at auction in 1899. A residential district now occupies the site of the post.

With the discovery of gold in the Black Hills of South Dakota in 1874, and the increasing presence of the railroad, Sidney became the hub for the shipment of supplies, equipment and thousands of gold- and-adventure seekers.

Most, if not all, Lincoln Highway towns in Iowa and Nebraska were indebted to the railroads but none more completely than Sidney. This town was totally the product of the Union Pacific Railroad, even to the design and street plans. Sidney was named for Sidney Dillon, president of the UP. The completion of the 2,000 foot bridge over the North Platte River in May of 1876 (near

Sidney as it looked to my father in 1915.

Sidney, Nebraska 1915

Landseekers, about 1912

Hopeful homesteaders line up in front of the Cleyburne Building, 1001 Illinois Street (Lincoln Highway). Charles Callahan erected this imposing building in 1908, costing between $12,000 and $15,000. There was an elegant flat on the second floor that was used by the Callahans.

today's Bridgeport) was crucial to Sidney's economy by helping to ensure a viable route north to the Black Hills. Sidney merchants loved it and supported it with their checkbooks. Tolls were charged for every man, vehicle or beast that crossed the bridge, including the military.

Sidney, along with other Nebraska communities, suffered the dust disaster of the 1930s. In fact, it was called the center of the "Nebraska Dust Bowl" with many farmers leaving for jobs in urban areas. Those farmers who stayed to fight the dust and wind worked with the state agriculture department to develop soil conservation methods, which allowed the ground to gather moisture during a fallow year and to prevent wind soil erosion. The rain and snow of 1940 ended the drought and Sidney is now known as the "Wheat Capital of Nebraska."

Today, Sidney is a community of 6,500 and a Lincoln Highway byway 1,654 miles west of New York and only four miles north of Interstate 80.

Treinen Garage, 1916

1040 Illinois Street (Lincoln Highway). l to r; Roy Green, John Treinen, Jack Bowman. John Treinen was an excellent auto mechanic and farmer.

Plowing

W. J. Davis, breaking Cheyenne County soil in early 1900s with a ten bottom plow. Today Sidney is known as the "Wheat Capital Of Nebraska."

Dirt Roads BY LEE PITTS

ADAPTED FROM *People Who Live At The End Of Dirt Roads*

What's mainly wrong with society today is that too many Dirt Roads have been paved.

There's not a problem in America today, crime, drugs, education, divorce, delinquency that wouldn't be remedied, if we just had more Dirt Roads, because Dirt Roads give character.

People that live at the end of Dirt Roads learn early on that life is a bumpy ride. That it can jar you right down to your teeth sometimes, but it's worth it, if at the end is home…a loving spouse, happy kids, and a dog.

We wouldn't have near the trouble with our educational system if our kids got their exercise walking a Dirt Road with other kids, from whom they learn to get along. There was less crime in the streets before they were paved. Criminals didn't walk two dusty miles to rob or rape, if they knew they'd be welcomed by five barking dogs and a double-barrel shotgun.

And there were no drive-by shootings.

Our values were better when our roads were worse! People did not worship their cars more than their kids, and motorists were more courteous, they didn't tailgate by riding the bumper, or the guy in front would choke you with dust & bust your windshield with rocks.

Dirt Roads taught patience. Dirt Roads were environmentally friendly. You didn't hop in your car for a quart of milk. You walked to the barn for your milk. For your mail you walked to the mailbox. What if it rained and the Dirt Road got washed out? That was the best part, then you stayed home and had some family time, roasted marshmallows and popped popcorn and pony rode on Daddy's shoulders and learned how to make prettier quilts than anybody.

At the end of Dirt Roads, you soon learned that bad words tasted like soap. Most paved roads lead to trouble. Dirt Roads more likely lead to a fishing creek or a swimming hole.

The only time we even locked our car was in August, because if we didn't some neighbor would fill it with too much zucchini.

At the end of a Dirt Road, there was always extra springtime income, from when city dudes would get stuck, you'd have to hitch up a team and pull them out. Usually you got a dollar…always you got a new friend… at the end of a Dirt Road!

Iowa State Highway Commission

Iowa Mud, *the Lincoln Highway, 1915, In its youth.*

WYOMING
And The Lincoln Highway

36 Lincoln Highway towns – 1915427 miles – 1924

"WYOMING IS A SMALL TOWN with a long Main Street" was the comment of Merrill G. "Sarge" Sargent of Pine Bluffs.

My impression of Wyoming has been "High Country" and wide open spaces, an impression confirmed on my first visit in 1968, and even later while traveling by truck, and still later while following my father's diary in June 2000 and 2001. The highest point (8,874 feet) on the Lincoln Highway is in, appropriately, Wyoming.

Also in Wyoming are the tracks of all major trails West: Oregon Trail, California Trail, Cherokee Trail, Mormon Trail and others. It is home to the world's first

Wyoming State Archives, Department of State Parks and Cultural Resources.

Tree In a Rock

Tree In a Rock

Lincoln Highway at the Tree in the Rock. A legend says that in the spring of 1868 when the Union Pacific Railroad was being constructed west of Cheyenne, the surveyors came across this struggling small tree growing out of a solid boulder of 1.43 billion year old Sherman granite. Legend also says that the original railroad tracks were diverted slightly to pass by the struggling tree. It also said that trains would stop here in dry seasons while locomotive engineers would "give the tree a drink."

The little limber pine has not grown much in the last 134 years. It is located south of 8,874 foot Sherman Summit, the highest point on the Lincoln Highway.

Lincoln Monument

Lincoln Monument, near Sherman Summit, 8,835 feet elevation. The Lincoln bust is 12.5 feet high weighing 3.5 tons. It sits on top of a 30-foot granite base. The monument was first unveiled on October 18, 1959, then moved to the present site in 1968 following the completion of I-80.

National Park—Yellowstone—with its incredible Old Faithful geyser. On September 6, 1870, Wyoming gave birth to Women's Suffrage with Mrs. Louisa Swain the first woman in the United States to cast a vote. In 1924 Nellie Tayloe Ross was the first elected woman governor. These women's issues gave Wyoming the nickname, "Equality State."

Wyoming entered the ranks of statehood on March 27, 1890, when President Benjamin Harrison signed the bill making it the 44th state. Wyoming's population is 494,000, or 5.1 persons per square mile.

There is evidence to indicate that Wyoming history reaches back to 12,000 years of prehistoric occupation by big game hunters, followed by the precursors of the historic Indians. Forty miles east of Lovell, Wyoming, is the Medicine Wheel, with twenty-eight spokes and a circumference of 245 feet—an ancient shrine built of stone by some forgotten tribe. This remains one of Wyoming's unsolved mysteries.

The occupation of Wyoming in modern times stems from the fondness of European gentlemen for beaver top hats, luring trappers from Spain, France and England.

Wyoming Tales & Trails

The Great Race

Entrant into the Great Race on the future Lincoln Highway, west of Cheyenne, Wyoming. It is accompanied by a Studebaker pilot car in foreground, March 1908. The Great Race was an unbelievable undertaking, especially in 1908. It was a race, believe it or not, around the world from New York to Paris. The race was won by a United States built Thomas Flyer in 169 days.

Stan Taggart

Church Butte

Church Butte provides a majestic backdrop for this Packard on the Lincoln Highway ten miles west of Granger

Heading West

This Packard is heading west on the Lincoln Highway between Rock Springs and Green River, Wyoming on its way from Johnstown, Pennsylvania to Salt Lake City.

Stan Taggart

Ames Monument

My father parks his Little Giant at the foot of the Ames Monument, on the eastern face below the medallion of Oakes Ames. The massive native pink granite monument, costing the Union Pacific Railroad $65,000 in 1882, is sixty feet high. Oliver Ames, brother to Oaks Ames, was the third president of the Union Pacific Railroad. The monument could be viewed from the Lincoln Highway until the 1930s when the route was moved farther south.

Nissley Family collection

Nissley family collection

Palisades

My father's Little Giant truck at the Palisades west of Green River, Wyoming.

Next came the growth of open-range cattlemen, followed close on the heels by homesteaders, railroads, and the building of towns.

Covered wagon trails were replaced by railroad tracks in the mid-1800s, which were ultimately replaced by the Lincoln Highway in 1913, which was replaced by I-80 in the mid-50s, bringing Wyoming's transportation history up to date.

My father commented on only seven Wyoming locations: Pine Bluffs, Burns, Sherman Hill, Carbon, Wamsutter, Granger, and Evanston. It seems that by now he was anxious to get on with pursuing the purpose of this trip—San Francisco, so from here on his photographic activities were curtailed.

Old Post

An early wooden Lincoln Highway post at the site of the Tree In A Rock.

Rock River, Wyoming

206 S. 4th Street at the home of Eva Mae Emerson, former postmaster of Rock River.

Medicine Bow, Wyoming

On the ground in front of the historic Virginian Hotel.

Rock River, Wyoming

Avenue C in front of the old hotel at the west end of town.

Medicine Bow, Wyoming

Across the road from the historic Virginian Hotel.

PINE BLUFFS, *Wyoming*

IN HIS OWN WORDS:

Tuesday, October 5, 1915

Cold. Traveled 60 miles to Pine Bluffs, Wyo. Booked & showed Victory. Had $21.60 House. Camped at Opera House.

Wednesday, October 6, 1915

Made up local view post cards for drug stores. Camped 8 miles west of Pine Bluffs.

PINE BLUFFS WAS ONCE the center of a vast hunting area for the Arapahoe, Cheyenne, Ute, Sioux, Blackfeet, and other Indian tribes. Even as late as 1850 census records show no white population in the area. The abundance of buffalo, black-tailed deer, elk, and antelope made the area popular with the more than 50,000 Indians of the region. This was part of the Great Plains of North America, which runs from inside Mexico north through the United States and into the prairie provinces of Canada. The United States portion, 1,300 miles long and 600 miles wide, has an elevation of 2,000 feet from the eastern fringe to 6,000 feet in the Rockies.

The second phase of Pine Bluffs history, an important link in the development of Wyoming and adjacent territories, was the arrival of the cowboys herding thousands of longhorns from Texas north as far as the Dakotas. The significance of this era of United States development, and the happenstance that the Lincoln Highway intersected this movement at Pine Bluffs, justifies a brief history of the Texas longhorn industry.

The Texas longhorns can be traced back to the Spanish explorers who in the sixteenth century introduced them to Mexico. During the Texas Revolution of 1835-1836, precursor to President James Polk's Mexican War, many Mexican ranchers retreated across the Rio Grande River, abandoning their ranches and thousands of head of Spanish longhorns which thrived and multiplied on the lush grazing lands of the Nueces Valley, the southernmost end of the Texas cattle range. The Anglo-American occupants declared all these unbranded cattle public property and many were rounded up by Texans and marked with their

Pine Bluffs, Wyoming

Pine Bluffs as seen from the pine bluffs east of town.

Pine Bluffs, 1915

E. G. "Mike" Sanders gave this postcard to his grand-niece, Anne Shields, Editor of Pine Bluffs Post. "Mike," six years old when my father came to town in 1915, found it among some old papers and thought it might be of interest to local residents. It is possible, but not likely, that this is one of my fathers photos. It is a view of Main Street looking south from a coal chute along the Lincoln Highway and Union Pacific Railroad.

Pine Bluffs, 1910

Looking north, circa 1910-1913. Note the Union Pacific depot and rail cars in center left.

own brands. A few unsuccessful attempts were made to market the cattle in New Orleans and Cuba, but it was not until after the Civil War and the growing city markets of the North, that the longhorns became a viable commercial product. Cattle worth three to six dollars in Texas were worth forty dollars per head in the North. The Anglo-American ranchers learned cattle raising from the Mexican cattlemen, and the cattle business from the vaqueros, even patterning their equipment after that worn by Spanish cowboys and adopting their terminology.

In the 1860s a few droves ventured into Indian territory and eastern Kansas where buyers from Iowa and points east were paying high prices. However, the Indian tribes and settlers along the way objected to the collateral damage done by the animals, including sickness from ticks carried by the rib-thin longhorns.

The shift of the Texas Trail to western Nebraska and southeastern Wyoming was the result of poor public relations between the Indians and Kansas settlers. The route of the trail through Pine Bluffs was largely due to the lush grass and abundance of clear water in the Lodgepole Valley near Pine Bluffs. The grass and water were welcomed by both cowboy and beast. From 1866 to the end of the century, Pine Bluffs was an important watering hole and shipping point along the Texas Trail. For several years more cattle were shipped from Pine Bluffs than any other place in the world. The last herd of 2,500 trailed into Pine Bluffs in 1899. On August 1, 1948, a fitting monument to the Texas Trail was dedicated in Pine Bluffs at the intersection of the Texas Trail and the Lincoln Highway.

Railroads dominated the third phase of Pine Bluffs history. Pine Bluffs is not much different from other Lincoln Highway towns except for the pine atop the pine bluffs. The pine trees were a ready source of wood for railroad ties and fuel. For a time it was a major industry. Crews of several hundred men chopped the trees, and using ox, mule, and horse teams hauled them to the tie camp where they were shaped into ties for the railroad. Twenty dollars a cord was paid to the choppers and the plentiful source yielded a good supply.

A huge blaze was fueled one day when 5,000 cords piled along the tracks caught fire.

The birth of the Union Pacific Railroad came about when Congress passed an act in 1862 authorizing construction of a trunk branch, "At a point on the western boundary of Iowa—in the territory of Nebraska." The railroad dream became reality when the first rail was laid on July 10, 1865 and the last spike driven on May 10, 1869

Merrill G. Sargent

Retired from the post office, "Sarge" is a long-time resident. He was very helpful with pictures and stories of Pine Bluffs. (Photo, June 19, 2001)

Pine Bluffs, Wyoming

The pine bluffs of Pine Bluffs, Wyoming. Note the Lincoln Highway in the foreground

at Promontory Summit, Utah. It was the railroad which was mostly responsible for putting Pine Bluffs on the map. Not so fast! There were serious conversations in California as early as May 19, 1856, when John B. Weller, senator from California, petitioned the United States Senate with a resolution to construct a wagon road from Missouri Valley to the Great Salt Lake basin, on the eastern slope of the Sierra Nevadas and on to the California border. California would take it from there, so as not to use federal funds for California roads. California argued that a wagon road was in the best interests of the nation by uniting the 500,000 California residents and the rest of the country. The *Sacramento Daily Union* said,

> From the East we must obtain our permanent agricultural population, and in order to give them a chance to come with their wives and children and their household goods, cattle, horses and sheep, a good emigrant road must be built across the continent and the immigrants protected from the murderous rifle and tomahawk of the Indian. The accomplishment of this national object the people of this state have a right to demand Congress.

The federal government responded by funding reconnaissance crews to survey for the improvement of wagon roads. Happily, the cartographer included the name of Pine Bluffs, thus putting Pine Bluffs on the map a good thirteen years before the first railroad engine puffed its way over rocking rails into Pine Bluffs. Sixty-six years later the Lincoln Highway received the vision for uniting the East and West with a wagon road, but by then wagons had installed engines and greatly multiplied their numbers far beyond Senator John B. Weller's imagination.

Tuesday, October 5, 1915, saw my father arrive in Pine Bluffs from Sidney, Nebraska. He booked a local theater (Opera House?) and made postcards for sale in local drugstores. Pine Bluffs was the last town he photographed on this trip west. He picked up his trade again in Omaha, Nebraska in late November on his return trip.

Pastime Theater

The Theater booked by my father, October 5, 1915.

BURNS, *Wyoming*

IN HIS OWN WORDS:

Tuesday, November 9, 1915

Cloudy, cold. Traveled 57 miles to Burns, Wyoming. Good roads.

Wednesday, November 10, 1915.

Cold. Left camp at Burns.

BURNS IS PROBABLY THE ONLY TOWN on the Lincoln Highway which came riding into history on auto wheels. The Federal Land and Securities Company was anxious to entice people to come to the "Golden Prairie" and purchase land which they had bought from the Union Pacific Railroad. They even bought six cars to take prospective buyers around the community. In addition to the six cars they were obliged to hire farmers to help with their horses and buggies. At one time there were up to 250 people in town with insufficient housing. The hotel was full, so some were obliged to bed down on the second floor of the bank building. The auto was an early blessing but also a later contributor to Burns' decline in the early 1920s when better cars and better roads made it easier for the motoring public to commute to Cheyenne, the self-styled "Magic City of the Plains." The Lincoln Highway, of course, was an active player in this drama providing a good road (my father said "good roads" in 1915) for the swarm of early prospectors as well as the road away from Burns during its decline.

Since four out of five early settlers were German Lutheran, the land company named the town Luther as an enticement to German immigrants, but the Union Pacific rejected the name because there was already a Luther, Wyoming. The railroad refused to build a depot unless the name was changed. The Lutheran Land Company gave in and the town was renamed Burns in 1911, to honor a Union Pacific employee killed during the construction of the tracks.

A human interest story emerged from early Burns, with a twist of romance, involving a Rev. Alfred Rehwinkel, a senior seminary student at Concordia College in St. Louis. Rev. Rehwinkel was in town for a student-preaching assignment and was on his way from an evening dinner to Burns after dark when his horse threw him into a barbed wire fence cutting his leg to the bone. He made it to the Burns hotel where Carl Kegler, owner/operator of the hotel, took him to a lady homesteader just south of town, Dr. Bessie Fell, whom Rehwinkel had met just that afternoon at the café in Burns. Helpful volunteers at the hotel had filled the wound with flour; it took Dr. Bessie two hours to clean the wound and stitch it up – all without anesthetic or a whimper. This chance encounter found its conclusion several years later at the altar when Dr. Bessie became Mrs. Rehwinkel. Alfred Rehwinkel went on to become a professor at the Concor-

Neg. #21279, Wyoming State Archives

Birdseye View

Burns, Wyoming as the birds would have seen it in 1915.

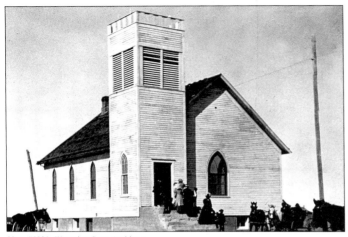
Presbyterian 1911

Burns Presbyterian Church dedicated March 5, 1911

Presbyterian 2001

Burns Presbyterian Church, June 19, 2001.

Lutheran 2001

Immanuel Lutheran Church, June 19, 2001. Dedicated June 13, 1919. Burns was formerly called "Luther," but was changed to "Burns" by the railroad who refused to build a depot unless it was named Burns.

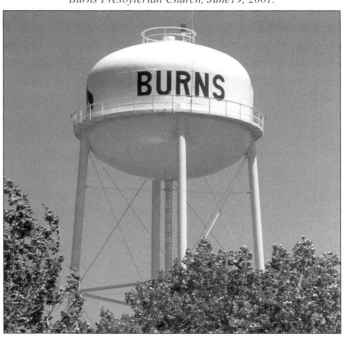
Water Tower

dia Seminary and in 1963 wrote *Dr. Bessie*, a book about his wife, who was the great-granddaughter, granddaughter, and daughter of doctors before her. She also had a brother and an uncle who were doctors. The book was in the making for several years but unfortunately didn't make publication until just after her death, "The greatest disappointment of my life," said Rev. Rehwinkel. They were married for fifty years. *Dr. Bessie* has become a collectors item.

Burns is still a comfortable farming town (3,731 feet elevation) on the north edge of the Lincoln Highway I-80 in Laramie County. It has a population of 1,303.

William Jennings Bryan

The day William Jennings Bryan spoke in Burns, Wyoming.

Neg. #14582, Wyoming State Archives

Steam Plow

Lux and Wheeler steam plow at Rider's Ranch, 1916. Dry land farmers had to hire big rigs to convert cattle ranches into farm land.

Neg. #5807, Wyoming State Archives

Burns, 1915

Burns, Wyoming, as my father would have seen it on October 5 & 6, 1915. The Lincoln Highway passes to the south of town.

Neg. #27483, Wyoming State Archives

CARBON, *Wyoming*

IN HIS OWN WORDS:

Thursday October 7, 1915

Cold. Very cold.
Water frozen 1 inch thick in buckets. Traveled 88 miles to Carbon, Wy. A deserted mining town.
Rough sheet trail all the way.

FOLLOWING MY FATHER'S DIARY on our first trip in June 2000, was a kind of seat-of-the-pants thing. My wife and I had a general idea where we wanted to go but specifics were unfolding surprises every day and appreciated mostly in retrospect. Carbon was like this. My father talked about Carbon as a deserted mining town but it was important enough to earn an entry in his diary. "Cold. Very cold. Ice 1 inch thick in buckets," he said, so we put Carbon on our list. Alas, it was not on our map nor was it even mentioned in any literature in our possession. A scruffy looking man with shifty eyes behind the counter of a small truck stop near Walcott, said he didn't know anything about Carbon either, but my disappointment or gentle insistence must have coaxed enough for him to re-remember. He told us to take Route 30 east to the first dirt road beyond the railroad tracks and turn right – this would take us to all that is left of Carbon—a cemetery. He added that once a year a few people would gather there. We knew his directions were correct when nineteen miles later we crossed the UP tracks, turned right on a dirt road, crossed the tracks again and saw a small weather-beaten, almost illegible sign tacked to a pole, with a little arrow pointing to "Old Carbon." This dirt road led off into the hills with no civilization in sight. It was after 5:00 P.M. and fear of the unknown persuaded us that the better part of valor was not to visit a strange cemetery in the empty hills of Wyoming in the evening.

It was a year later, that we finally visited Carbon. Our first stop was at the Chace Ranch where Kaylyn Chace graciously gave us permission to proceed. About five miles and two gates later we rounded a curve and before us stood four spruce trees, sentinels of time, keeping

Carbon Cemetery

Carbon cemetery, where a town of 4,000 buried their dead—many from mine catastrophies.

Four Sentinels

Four cedar sentinels stand watch over the Carbon cemetery.

watch over memories, at the base of the cemetery hill. It is a large cemetery taken over by natures' sagebrush and surrounded by a post fence. A wide variety of names like Grabenhorst and Simelius were recorded on tombstones, several of which were of recent origin and well cared for. Obviously there are descendants of Carbon who still remember and care.

Around the next bend were the remnants of Carbon. No buildings. We had to venture off the road into the brush to find crumbling stone foundation walls. Someone had erected a sign identifying the site of the Carbon State Bank, and another pointing to Carbon's Chinatown and train station a half mile east. Carbon, fourteen miles west of Medicine Bow, began in 1867 with the discovery of coal by a Union Pacific Railroad geologist. The town was built a year later for the sole purpose of providing coal for Union Pacific steam engines. At its peak seven mines were in operation, with some crews removing up to 100 tons of coal per day. By 1890 Carbon reached its peak population of around 4,000 but in the same year the fortunes of Carbon began to fade, first by a serious fire on June 19, 1890, which was stopped only by dynamiting some buildings, and secondly by the relocation of the railroad to Hanna to avoid the six-mile Simpson Hill which required

Road To Carbon

Lincoln Highway to Carbon, Wyoming. Carbon is five miles south of the present Lincoln (Route 30). My father traveled this road or close to it on Friday October 7, 1915.

Carbon Bank

All that remains of the Carbon State Bank, organized December 9, 1881

double-heading the locomotives. After thirty-four years of producing quality coal, Carbon expired in 1902, when the last of the big mines played out.

The story of coal in the region goes back to the Cretaceous period, some 140 million years ago. It was discovered in Wyoming by the Fremont expedition of 1843 and underlies forty percent of the state of Wyoming, the one-time largest producer of coal in the United States. The nearest competitor to Carbon was Hanna (also on the Lincoln Highway), only ten miles away and established in 1886, nineteen years after Carbon. On June 30, 1903, an explosion in Hanna Mine #1 killed 171 miners, rendering 600 children fatherless, the worst disaster in the state. Another explosion in 1908 at the same mine killed fifty-eight miners.

Eventually the advent of diesel fuel supplanted the need for coal to fire steam engines, which put coal towns like Hanna on hard times. Today there are rumors that the mines may be reactivated, not to feed steam engines but as a cheap source of energy for power plants and other industrial uses.

Westward

Present Lincoln Highway threads its way westward through Carbon County.

WAMSUTTER, Wyoming

IN HIS OWN WORDS:

Saturday, October 9, 1915

Warm. Broke camp 7:30 Traveled 84 miles to Wamsutter, Wyoming. Rough day. Alkali roads through sheet country.

Sunday, October 10, 1915

Cold. Broke camp AT 2:00 a.m.
6 miles east of Wamsutter.
20 miles out broke aux. Spring.

JUNE 1, 1870, IS AS FAR BACK AS AVAILABLE records go to the first evidence of life in "Washakie," a station between Latham and Red Desert with its umbilical cord still attached to the Union Pacific Railroad. Somewhere between July 1885 and September 1886, and for some unrecorded reason, the name on railroad passenger reports was changed to "Wamsutter." Adult one way railroad fares in 1915 from Wamsutter to:

RED DESERT	$0.30
ROCK SPRINGS	$2.35
LARAMIE	$4.75
OMAHA	$18.19

Children under 12 and clergymen were half-fare.

As noted often before, most Lincoln Highway towns were beholden to the railroads for commerce, travel or whatever, until 1913 when the Lincoln Highway liberated them and opened a vast new world of opportunity. Even then, the railroads continued life support to those developing communities. The fine sheep-grazing land of

Conoco Station *Original Conoco filling station*

the Red Desert provided the economic source for wool, Wamsutter provided the location and the Union Pacific Railroad provided the means for marketing.

Wamsutter was once one of the largest wool shipping centers in the United States with over a million sheep annually wintering in the area. Thirty-one different sheep companies operated out of Wamsutter. This industry continued, albeit in a gradual decline until the 1930s, when oil and its companions supplanted agriculture. Today Wamsutter is a natural gas and oil town with oil trucks, drilling rigs and an oil culture dominating the streets and town life. In 1938 the Standard Pipe Line Company built an eight-inch line from Fort Laramie to Salt Lake City to transport crude oil from the Wyoming oil fields around Casper, Rock River, and Medicine Bow to refineries in Salt Lake City. Wamsutter hosted one of the three pumping stations which originally served as backup and later was brought into full service as

Remnant of old Wamsutter school.

the tracks, finishing it in 1910. The first school opened in January 1904, with Andy Bugas as treasurer. In 1916, Andrew P. Bugas became both postmaster and mayor in the same year. Years later, one John Bugas, a son of Wamsutter, retired from the Ford Motor Company and became famous for his philanthropy, and in 1914 the city park was created under the leadership and tireless efforts of the "driving force" in Wamsutter, Grace Bugas. She was 63 when murdered by two young men whom she had befriended. The park was named, "The Grace Bugas Memorial Park". Not the least of Bugas contributions to Wamsutter was the information about the town found in the 1924 Lincoln Highway Guide provided by (who

more volume and heavier oil was needed, especially for the demands of World War II. At that point Wamsutter became a major player in the oil game.

The history of Wamsutter repeats the names of certain persons who provided the backbone, stability and vision necessary for an isolated western town struggling for its existence. Bugas was one of those pioneer names. Andy P. Bugas along with a Mr. Harnish built the first saloon in Wamsutter in 1906 on the north side of the tracks, but not without protest from the Woolgrowers

Intersection

Lincoln Highway intersection in Wamsutter, Wyoming. Note the old school on the left and the water tower on the right.

else?) Lincoln Highway Consul Andrew P. Bugas.

When my wife and I visited Wamsutter in June 2001, our inquiries led to another longstanding Wamsutter name —Waldner. Verne Ardell and Emma Waldner came to town in 1956 from Freeman, South Dakota. They bought the Conoco station from Gerald and Mid Gleason and thus planted their family roots in Wamsutter. We learned to know Ken Waldner, the third son of Verne and Emma Waldner, who now operates the Conoco station. Ken went out of his way to be helpful, even providing a copy of the *Wamsutter Centennial Book—1890-1990.*

Wamsutter has never aspired to the big city stuff. It is still a small town plugging away at being itself on the Lincoln Highway. It even had a Lincoln concrete post until it fell victim to a train wreck and disappeared in the cleanup.

Watertower

ginal Wamsutter water tower purchased from the Union Pacific Railroad

Association, who petitioned the county commissioners of Sweetwater County to protest a saloon in Wamsutter. In 1908 Andy Bugas began building the first hotel, south of

EVANSTON, *Wyoming*

IN HIS OWN WORDS:

Wednesday, November 3, 1915

Fair. Left Summit early at 6 a.m.
Traveled 6 miles east of Evanston. Made 66 miles.
Had trouble with motor getting hot on account
of carbonizing. Camped six 6 miles east

Thursday, November 4, 1915.

Fair. Took motor apart and removed carbon from
cylinder heads and put it back. Left camp at 2:30.
Traveled 71 miles; motor OK.

AS WE TRAVELED THE LINCOLN HIGHWAY West from eastern Pennsylvania, the diverse ethnic composition of Lincoln Highway towns was impressive. In Wyoming Chinese joined the mix, first as temporary laborers building railroads and later as settled coal miners and shopkeepers. In spite of their industry and hard work they failed to escape the unfortunate brunt of senseless, cruel discrimination.

One of the prominent early landmarks of Evanston was Chinatown, occupied, some say, by the largest concentration of Chinese outside of California. They worked for the Union Pacific Railroad and coal mines at nearby Almy. The Chinese in Evanston began in Rock Springs, where they worked for the railroad and in coal mines, and where their story contributes no pride to American history. In September 1885, festering resentment climaxed in the belief that the Chinese kept wages down and received favorable work assignments. The result was the massacre of many Chinese and the burning of their homes. Many of these Rock Springs Chinese fled to Evanston where they lived in tar-papered shanties from the late 1870s to the early 1920s. It could be argued that the most important Chinese contribution was the highly decorated temple called the Joss House. On January 16, 1922, they were evicted from their Union Pacific-owned homes, and forty-five minutes later the Joss House burned to the ground. To the credit of modern Evanston, the community recognized the important legacy left by

Stan Taggert

Payson Spaulding

Payson W. Spaulding (second from right) stands in front of his Evanston law office, opened in 1901. He owned the first automobile in the county (Studebaker?) so he could drive the emerging Lincoln Highway, for which he became State Consul in 1913.

Stan Taggert

Coast to Coast

This photograph is one of several found in Payson W. Spaulding's home after his death. The group is supposedly the first family to make the coast-to-coast trip from west to east in 1908. Mr. Spaulding is probably in the driver's seat. He was also the Lincoln Highway Consul.

the Chinese by building a replica of the Joss House in 1990 to preserve the memory of their rich heritage.

Evanston began as many other Lincoln Highway towns, with one man and the railroad. In November 1868, as the grading crews of the Union Pacific approached the site of Evanston, Harvey Booth pitched a tent near what is now Front Street and opened a saloon and restaurant. That tent, with wooden floor and canvas sides, is now recognized as Evanston's first building.

On December 1, 1868, the railroad reached Evanston where a depot was built one year later, in 1869. The town was plotted by James A. Evans, the Union Pacific Railroad's surveyor, for whom the town was named—Evanston.

Waldemar Anderson (1893-1972) was one of many persons important to Evanston and to the Lincoln Highway.

As a successful local businessman, he founded the Hotel Waldemar in addition to assuming, over the years, many other civic responsibilities. At least two events endear him to the Lincoln Highway. He was an Evanston patriot who graduated from the Evanston High School on May 29, 1913, the birth year of the Lincoln Highway, and he lived most of his life on the Lincoln Highway.

Stan Taggart is another Evanston resident of note, perhaps the most relevant of all to the Lincoln. At Exit 108 on I-84 in Utah, the arrow points to "Taggart," where Stan grew up—on the Lincoln Highway. It was here his father and uncle owned a Conoco station, cabins, and other things to entice tourists. Today Stan is proprietor of Rock 'n' Rye, Pete's Bar and Saloon. It seemed that everyone in Evanston with whom we talked smiled pleasantly at the mention of Pete's Bar and Stan Taggart, an indication of warm feelings and good memories. It is on the old Lincoln Highway about one mile east of town just off I-80, Exit #6. My wife and I had the pleasure of visiting Stan on location in Pete's Rock 'n' Rye. Stan is an engaging guy who went far out of his way to be helpful, providing information long after our visit.

Evanston today is an active, upbeat community in the land where the Bear River flows; where everyone, it seems is ready to sing the praises of the town at the drop of a hat. The town has a population of 11,500 and an annual 300 days of sunshine.

Stan Taggert

Anderson House

Waldemar Anderson stands in front of the house, on the Lincoln Highway, in Evanston, Wyoming, where he lived most of his life.

Rock 'n' Rye

Pete's Rock 'n' Rye Saloon One mile east of Evanston, Wyoming, on the Lincoln Highway.

Stan Taggart

Stan Taggart stands before one of the several gables of his Rock "n" Rye Saloon

You Auto Know

Birth year of the Lincoln Highway—1913 *(page 3)*

Length of the Lincoln Highway—3,389 miles *(page 3)*

Number of Lincoln Highway states—13 *(page 44)*

First president of the Lincoln Highway Association—Henry B. Joy *(page 3)*

Official car of the Lincoln Highway Association—Packard *(page 5)*

First seedling mile—Malta, Illinois (page 73)

Significance of the Panama-Pacific International Exposition—celebration of the completion of the Panama Canal *(page 1)*

First car to travel the Lincoln Highway—Saxon *(page 10)*

Location of the famous Lincoln Highway coffee pot—Bedford, Pa. *(page 41)*

Date of "planting" the original concrete marker posts—September 1, 1928 *(page 11)*

Date named highways were replaced with numbers—1925 *(page 4)*

First vice president of the Lincoln Highway Association—Carl G. Fisher *(page 65)*

Last field secretaery of the Lincoln Highway Association—Gael Hoag *(page 4)*

Date of the rebirth of the LH Association—1992 *(page 4)*

Most famous LH bridge in Iowa—Tama, Iowa *(page 84)*

Western terminus of the LH—Lincoln Park, San Francisco, California, *(page 201)*

Manufacturer of the original 1928 concrete posts—nobody knows

The burning of what LH bridge affected the Civil War?—Columbia's 2nd bridge *(page 17)*

Estimated cost to pave the LH coast to coast in 1913—$10 million *(page 3)*

Lincoln Highway Association headquarters—Franklin Grove, Ill. *(page 73)*

Most famous LH gas station—George Preston's, Belle Plaine, Iowa *(page 92)*

Total number of original concrete marker posts—2,346 *(page 11)*

First Indianapolis 500—1911 (page 66)

Packard slogan—"Ask The Man Who Owns One" *(page 61)*

UTAH *And The Lincoln Highway*

232 miles – 1924   27 towns/ranches - 1915

IN 1847 WHEN THE MORMONS SETTLED into the Great Salt Lake Valley, the western boundary of the United States was the Missouri River. The Great Salt Lake Valley had belonged to Mexico, the result of Mexico's fight for freedom from Spain in 1822. The land became United States property in 1848 under the treaty of Guadalupe-Hidalgo following President James Polk's self-initiated war against Mexico for the land which now comprises Calif., Nev., Utah, plus a part of N. Mex., Ariz., Colo., & Wyo.

Given the earlier persecution of Mormons in Illinois and elsewhere, plus the feeling that this land was up for grabs, it is not hard to understand Brigham Young's desire to make Utah a separate state with himself as governor. The federal government took a dim view of this idea and forthwith dispatched a military force of 2,500 men to convince Young, by force if necessary, that the status of Utah would be otherwise. Thus occurred what was called the "Mormon War," in which the Mormons claimed no casualties, a claim disputed by others. In any case, the military in 1858, under Albert Sidney Johnston, reached a compromise with Brigham Young. The Mormons permitted a non-Mormon governor to be seated. Statehood for Utah was delayed by Congress until Mormon Church President Wilford Woodruff announced the abandonment of polygamy in 1890.

The Lincoln Highway across Utah was not without its share of fireworks. The competition was not between local communities, as in the eastern states, because Utah had only vast expanses of desert inhabited by widely scattered ranches. Competition came from southern California (Los Angeles), northern California (San Francisco), the Lincoln Highway Association and Utah state bureaucracy. Los Angeles favored a southern route—somewhat attractive to Utah as well because it kept the tourists and their money in the state longer—San Francisco pressed for a northern route just south of the Great Salt Lake. The Lincoln Highway Association argued for a route through

Cabin

Donner-Reed Pioneer Museum cabin Grantsville, Utah.

Jail

Jo Ann Hanson stands beside the jail at the Donner-Reed Pioneer Museum Grantsville, Utah.

Special Collections, University of Michigan Library

Orr's Ranch

Goodyear's Wingfoot Express enjoys a respite at Orr's Ranch in 1918. Goodyear pioneered the first practical pneumatic tires for heavy trucks. Orr's Ranch, seventy-one miles southwest of Salt Lake City, was right up there with the best in providing critical survival services at a critical place on the Lincoln Highway. The Orr brothers (Mrs. William Orr too) provided a good dinner for 75cents with 50 cents paying for a space to pitch a tent, plus gasoline, radiator water and shade from the grove of poplar trees.

the Great Salt Lake Desert to Ely, Nevada. The Ely route was to a degree a compromise between San Francisco and Los Angeles because the mileage from Ely to either was about the same.

When my father made his trek in 1915 he followed an early route via Fish Springs, Ibapah, and Ely. Three years later in 1918 the Lincoln Highway Association came to an agreement with the State of Utah, donated $100,000 and signed a contract for a shortened path across the narrow end of the Great Salt Lake Desert to Gold Hill, Ibapah, and Ely. This was called the "Goodyear Cutoff" because Goodyear Tire and Rubber Company donated $100,000 toward it. The State of Utah accepted the money and partially honored the contract during 1919, then abandoned it in 1920 in favor of a route straight west of Salt Lake to Wendover, effectively isolating Gold Hill and Ibapah. In 1930 the Lincoln Highway officially went south from Wendover to Ely. Eventually even Ely was bypassed with the construction of I-80, not reconnecting with the historic Lincoln Highway until east of Sparks in western Nevada.

Utah, for my wife and me, was one of those never-a-dull-moment experiences. We entered Utah just west of Evanston, Wyoming, drove past the Taggart exit on I-84

Mountain Pass

Looking east from one of the many Utah Mountain passes.

Split

Miriam studies the Bureau of Land Management sign along the Pony Express Route.. My father likely did not pass this intersection

and entered Salt Lake City from the north. Later that day we enjoyed a delightful tour of Salt Lake City guided by a close relative who lives on location. At 6:00 A.M. the next day we fueled up in Salt Lake City and went west to Grantsville, where Jo Ann Hanson provided a tour of the Donner-Reed Pioneer Museum. Among other early pioneer memorabilia was a wounded but proud Lincoln highway concrete post—evidence that the Lincoln Highway was in fact here to encourage travelers across this desert, over the mountains and on to their western destination as far west as they wanted to go. In 1915 my father came through this area on Tuesday October 12, and camped "West of Grantsville in edge of the desert. No trouble."

Grantsville shares another place in history with the famous Donner party. In 1846, They stopped here for water a year before Brigham Young became a tenant in Salt Lake City. Their route through the Grantsville site was adopted 67 years later by the Lincoln Highway. The Donner party called it "Twenty Wells", a virtue recognized by later pioneers who were attracted by the streams and lush growth of willows.

Grantsville also suffered near starvation in 1855-56

Simpson Springs

Simpson Springs Station was named for J. H. Simpson, who stopped here in 1858. It is one of the most dependable watering points in this desert region. It was used by the Pony Express and the Overland Express.

Through The Desert

The Lincoln Highway snakes its way west south of the Great Salt Lake Desert.

at the mercy of the famous grasshopper infestation. The insects came in waves devouring everything that grew. Some farmers tried squashing them with large rollers, and burning them but they still kept coming and many people suffered terribly from lack of food and the devastation of crops.

After Grantsville we followed what we thought was my father's route, south through Tooele to Vernon where we headed west on the Pony Express/Lincoln Highway toward Fish Springs. Later it became evident that this was not likely my father's route, nor was this in fact the Lincoln Highway. We connected with the 1913 Lincoln about nine miles east of Fish Springs. For many miles beyond Vernon the gravel road was strewn with millions (I'm sure) of large grasshoppers and at Simpson Springs the floor and walls of the rest rooms were black with them. Are they possibly descendants of the Mormon grasshopper plague of over a century before? Someone later did indeed identify them as "Mormon Crickets."

As the road wound into the mountains, it became an experience in stone, sand, washboard, dust, and fantastic beauty. The vistas with their multi-colored hues of

Great Salt Lake Desert

Part of the Great Salt Lake Desert near Fish Springs, Utah.

Fish Springs

The spring at Fish Springs, Utah

the desert, viewed from the many passes, can only be described by sight. Neither words, brush, paint, or canvas are up to the task. We were later entertained by two separate groups of pronghorn and many fleet-of-wing birds, which we accused of being western kingbirds. The connection between the Lincoln Highway and ancient times is a fascinating observation. Even animals are not exempt. According to *Desert USA,* the pronghorns we saw have some pretty old ancestors, "Entirely unique on this planet, the pronghorn's scientific name, *Antilocapra Americana* means 'American antelope goat.' But the deer-like pronghorn is neither antelope nor goat—it is the sole surviving member of an ancient family dating back 20 million years."

By noon we arrived at wonderful Fish Springs—an oasis by any definition—on the south edge of the Great

Callao and Ibapah

Lincoln Highway between Callao and Ibapah, Utah.

Callao Sign

Callao sign points to Ibapah, 36 miles on the Lincoln Highway. In this area the Pony Express and Lincoln Highway travel together.

Salt Lake Desert. We learned that this is the site of the former famous John Thomas Ranch, where many motorists found refreshment and radiator water. Drake Hokanson suggested that "very few who ever passed this way forgot him (Thomas) or failed to mention him in their writings." My father must have been one of the "few" because he never said a word about the "six feet four inches tall and four feet wide" John Thomas, legend of Fish Springs, but then my father never commented on anyone he met on this trip. Fish Springs is now home to at least seventeen different species of waterfowl (by "home" meaning they nest here), and home-away-from-home for several times that many for birds other than waterfowl. Jay Banta, Fish Springs National Wildlife Refuge manager and Utah Director on the Lincoln Highway board, was most cordial and helpful, sharing important tidbits of the Lincoln to come. In fact it is safe to say that Jay shares some responsibility for this volume coming to fruition. He planted the original seed by suggesting that I write a report of our trip. Imagination and presumption did the rest.

On Wednesday, October 13, 1915, my father, "traveled all day in sandy desert. Made 106 miles. Camped at old well between Fish Springs and Kearney's Ranch." We found Fish Springs but where was Kearney's Ranch? Jay Banta said it is near Callao, a "town" about twenty-five miles farther on. He also said there are remnants of the old Kearney Hotel in Callao. When my father drove thru here in 1915 it was an important food-gas-rest stop for California-bound travelers. The day we arrived was very windy and dusty. We blinked and almost missed Callao. I guess we were looking for a "City Limits" sign. What caught our eye was a rather nice-looking building with a sign which read, "Callao School." There were a few other scattered ranch buildings but no sign of life—no people, cattle, horses—only tall weeds and dead grass blowing in the wind, a few gnarled survivor trees, a few pieces of very used farm machinery, and a discarded Gateway computer box. Finally a man emerged from a mobile home and kindly pointed to the remains of the hotel and the extent of the once elegant original facility. He said his dream is to restore it.

Our next watering hole was Ibapah, another twenty-six miles beyond Callao. At one time Ibapah was a major rest stop for weary Lincoln Highway travelers, where advertising promised: "Good meals, and home cooking at reasonable rates. Gasoline, oil and supplies. Owen Sheridan, Proprietor." As late as 1926, limo-sized Pierce Arrow buses made scheduled stops at Ibapah. This changed in the early '20s when Utah reneged on its commitment to build the Goodyear Cutoff and the route was moved to the Wendover Cutoff, some fifty miles to the north—now I-80. Since then Ibapah has become an isolated ranching

Kearney's Ranch

Remnant of the Kearney's Ranch Hotel in Callao, Utah

Callao School

Tintic School District, Callao, Utah. It was fifteen years old when my father stopped here in 1915.

community. We had been informed that gasoline was available in Ibapah but all we found were two very old padlocked gas pumps behind a chain link fence, which hadn't felt the pulse of gasoline in their hoses for many a moon.

One year later in June 2001, we revisited Ibapah and discovered that in fact, gasoline was available—at the site of the original Sheridan Hotel. It was accessed by a long lane from the main road. We bought some snacks and filled up with gas ($1.76 per gallon). The road through Ibapah was hard-surfaced blacktop, which veered off to the left at the south edge of "town," headed for the Goshute Indian Reservation. The graveled Lincoln proceeded straight ahead a short distance before heading westward into Nevada, the border being a line like the magical equator, a line we never saw.

Callao

Lincoln Highway between Callao and Ibapah, Utah.

Jay Banta

Jay Banta stands beside a Lincoln highway concrete post donated to the preserve. It came from somewhere in California.

Pioneer Museum

1928 Lincoln Highway concrete post inside the Donner-Reed Pioneer Museum, Grantsville, Utah.

SALT LAKE CITY, *Utah*

IN HIS OWN WORDS:

Tuesday, October 12, 1915

Broke camp at 3:AM. Run into Salt Lake City. Spent three hours.

SALT LAKE CITY CONJURES VISIONS of tall buildings with religious intent, temples, great music, clean streets, sea gulls, neat productive farms, long lines of marching burdened people pushing carts, ecclesiastical hierarchy and large families. The Mormons have experienced all the above plus internal strife and external persecution. Under the leadership of Brigham Young these people marched over the Wasatch mountains to the fertile valley of the Great Salt Lake on July 24, 1847. The party of 148 included 143 men, three women and two children. Settlement of outlying areas began immediately so that three years later, in 1850, Bountiful, Farmington, Ogden, Tooele, Provo. and Manti were settled. Immigration had swelled the population to 11,380. Basic industries developed rapidly but without hotels because no one was traveling, no barbershops

Salt Lake City

Salt Lake City Main Street in August, 1913.

Utah State Historical Society

because every man chose to shave himself or to grow a beard, no stores because no one had goods to sell, no signposts, beer shops, grocery stores, dry goods, or hardware stores. Even given that fifty percent lived on farms, it's still a lifestyle hard to comprehend, but it helps if we realize that life for them centered on the day's work and church activities. Here was a group of people equipped by experience to tame a harsh land and their communal religious faith underscored the necessity for cooperative effort.

Geologically, Salt Lake City goes back some 60 million years when the Rocky Mountains were being formed. This uplifting of the earth created two main ridges, the Wasatch on the east and the Sierra Nevada on the west. Between these two mountain ranges, the Great Basin and the Great Salt Lake eventually were formed – host to Salt Lake City.

This is the geological, social and political environment for the intrusion of the Lincoln Highway in 1913. The Lincoln was received with both controversy and genuine enthusiasm. The Rev. Frank G. Brainnerd of the First Congregational Church in Salt Lake City shared his view of the Lincoln Highway with his congregation on October 15, 1913:

State Capitol

Utah State Capitol, June 1915. Any or all of the many Lincoln Highway travelers to the Panama-Pacific International Exposition in San Francisco could have seen this building under construction.

Utah State Historical Society

The words of Reverend Frank G. Brainerd, October 15, 1913

"*It is a name to conjure by*", he said from the pulpit. "*It calls to the heroic. It unrolls a mighty panorama of fields and woodlands; of humble cabins and triumphant farm homes and cattle on a thousand hills; burrowing mines and smoking factories; winding brooks, commerce-laden rivers and horizon lost oceans.*

And because it binds together all these wonders and sweeps forward till it touches the end of earth and the beginning of sea it is to be named the 'Lincoln Highway.' It brings back to us the lank figure of the growing boy walking the country roadway with borrowed books; the reaming out, surveying and building of his highway of the soul, that should stretch from that mysterious ocean of the past, whence he came, to the mysterious ocean of the eternal, to which he would go; a highway long whose every day travel he had a gentle word for the sorrowing, and for the one in trouble, a sharp prod for the indifferent, a word of council for the perplexed, an inspiration for the doubtful, and love for all; the highway of the soul of the 'Great American.'"

Salt Lake

The Great Salt Lake as seen from the Lincoln Highway. At one time the lake was a fresh water inland sea over 1,200 feet deep with an outlet to the Pacific Ocean via Snake River. Evaporation has gradually reduced its size. With no longer an outlet, the salt content has risen to twenty-five percent; The oceans average only four percent.

Utah State Historical Society

Brigham Young

NEVADA
And The Lincoln Highway

38 Towns/Ranches - 1915   443 miles– 1924

ABOUT THIRTY MILES SOUTH WEST of Ibapah, Utah, and a tad past Tippett, Nevada, a sign pointed up the mountain to Antelope Range, Pony Express and Rock Springs Pass. Since we were in the mood for adventure and mistakenly believed that the Pony Express and the Lincoln Highway were fellow travelers, we took on the mountain, pointing our car toward the sky and Rock Springs Pass. The road soon narrowed to two tracks, sprouting tall grass in the middle. As bushes and trees encroached more and more from either side and the elevation became steeper, I ventured a comment: "At least the road can't get any worse." Famous last words! Shortly thereafter, picking our way around the biggest rocks and choosing carefully the shallower washouts, a truant rock broke the bracket on the left wheel suspension of our small camper, dropping that side to less than two inches from the road. Close inspection showed the trailer would not rub the tire, so we continued creeping onward and upward, even detouring around a fallen tree.

A few minutes later our mood for adventure was challenged by a small yellow light on the dash. The legend said, "Low fuel." We had no idea how far it was to the top and even less what to do if and when we got there. Another soul had not been seen for hours, and there was still the nagging fear that we might actually be on the wrong road, and come face to face with a dead end, and ultimately be gobbled up by the mountain. Our only consolation was an early-on small sign which read, "Pony Express Trail, 1860-1861." The Pony Express would certainly have open ends on each side of the mountain, and certainly someone travels it. Right?

As with most things we worry about but never happen, this was no exception. The mountain crest did finally show. The road even widened a bit as it rolled over the ridge, giving us space to exhale before plunging down

Rock Springs Road

The road to Rock Springs Pass starts here. Antelope Range is up ahead.

Rock Springs Pass

From the top of Rock Springs Pass it's all downhill—a good thing when the low fuel light glows.

the other side. There was a well-kept sign which read, "ROCK SPRINGS PASS, ELEVATION 7,890 FEET." About twenty-five miles and twenty shorter fingernails later, we spied in the distance what looked like a ranch. We were greeted by doors flopping in the wind, empty farm buildings and scattered pieces of old machinery. Soon after another ranch showed up. We thought this was also vacant until a woman's voice called to us.

"I have two questions," I said.

"How far is it to Schellbourne Ranch?"

"About seventeen miles" she said.

Rock Springs Elevation

Rock Springs Pass and the road just traveled.

My second question:

"Can I buy some gas?"

She said she would have to ask her husband who soon appeared, a tall Paul Bunyan built man in boots, jeans and black mustache.

"Can I buy some gas?" I inquired.

"I can't do that. It's illegal," he said.

"Then can I give you a donation?" I asked.

"How much do you have?"

"Much what?"

"How much gas do you have?"

I told him I didn't know because the low fuel light had been on for the past hour and a half. We talked a bit about our trip—where we were from and where we were going.

We also discovered that we were at Spring Valley Ranch. He saw the crippled camper, almost dragging the ground, and directed me to pull up to two large drums on stilts and asked,

"How much does it hold?"

"About fifteen," I said.

He proceeded to fill it up. I handed him a $20 bill. He smiled,

"You know it costs me over two dollars a gallon just to get it out here."

I handed him another twenty. He said,

"That's too much, give me another five and we'll call it square."

A sequel to this story occurred one year later, in June 2001, when again Miriam and I followed my father's diary. This time we did not go over Antelope Mountain at Rock Springs Pass. We went around the south end of the range by way of the "Block House," arriving at Spring Valley Ranch from the south. We spent about an hour and a half visiting with the rancher, Hank Vogler, who had so graciously come to our rescue the year before, and with his wife, Dana, and son, Stenson. We learned that Hank operates an extensive sheep ranch, hiring sheep herders from Peru. He also showed off his white Pyrenees sheepdogs which he unilaterally introduced to the local indus-

Hank Vogler

Hank Vogler stands beneath his gasoline storage drums at Spring Valley.

try, against the advice and dire warnings of his fellow sheep ranchers. However, his dogs did so well in bonding with the sheep that most ranchers now have followed suit. Hank also gave us a copy of the children's book he wrote about raising sheep. What wonderful people live on the Lincoln Highway!

Following our initial visit at Spring Valley Ranch in 2000, we limped the next fifty-five miles, pulling our crippled trailer into Ely, Nevada, and thoroughly enjoyed

Sheep Dog

Hank Vogler's Great Pyrenees sheepdog watches his sheep along the Lincoln Highway at Spring Valley, Nevada.

Schellbourne

Entrance to Schellbourne Ranch, a former Pony Express station.

Gary's Garage

Gary's machine shop in Ely, Nevada

Main Street

Eureka's Main Street, looking west.

a good night's rest in a fine motel. The next morning we found a walk-in clinic for sick and ailing campers. "Doctor" Gary North, from Gary's Machine Shop, performed the necessary "surgery." The initial examination revealed more than a broken suspension bracket—the whole rear section of the camper body had come loose from the frame. Gary, his son and staff did a thorough repair, but gave us much more—some modern Pony Express history. There is a local group with ten horses who once a year run the original 100-mile Pony Express trail between Robin, Nevada and Ibapah, Utah, carrying real mail. We found it interesting that one of their riders recently got lost going over Rock Springs Pass.

The late B. L. (Babe) Anderson from Cedar Rapids, Iowa, provides a Lincoln Highway voice from Ely past. Anderson grew up on a ranch on the Lincoln Highway near Ely, Nevada, and was 10 years old when the seventy-two-vehicle army convoy came to town in 1919. His memory of that event was one of his most graphic childhood memories. "After dinner the Goodyear band that was accompanying the Army came out in full uniform and played music for us for about two-and-one-half hours. I can remember sitting on the ground in front of the band with my mother; it was a delight," he said. After the band finished playing, one of the soldiers who had a pet raccoon asked Babe if he had a chicken he could have.

"We'd have given them a horse if they asked for it," Anderson said.

"When I got back with the chicken, he was holding the raccoon on his lap. It reached out with its paws and grabbed the chicken around the neck. When the chicken opened its mouth to squawk, the raccoon grabbed ahold of its tongue, yanked it out and ate it."

"Well, I thought that was about the most terrible thing I'd ever seen. I ran back to my mother petrified and shaking." Like a bad dream, not every memory on the Lincoln Highway has grace. In 1927 Anderson moved to Iowa, where he became head of his own asphalt paving company. He said that he became interested in road work while helping his father maintain an eighty-mile section of the Lincoln Highway, using a road grader and a team of horses.

From Ely we headed west on U.S. 50 across the Nevada desert. In 1915 this was the Lincoln Highway. Several persons warned us about the "Loneliest road in America" and said to be sure to have plenty of water, gas, and a death wish. "There is nothing, nothing on that deserted Nevada road" we were told. However, having spent the previous day in the dust of the deserted dirt Lincoln Highway in Utah, this stretch of blacktop in Nevada was beautiful and we covered it without a hitch. It was a different story my father told in 1915. According to his diary, the road was "sandy" and had "chuck holes." He got stuck in sand outside Fallon, had to clean the carburetor west of Ely, broke a spring just west of Fallon, took a wrong turn east of Austin and got lost in the desert. He got stuck in the mud flats and at 10:30 P.M. on October 16, 14 miles east of Austin, ran out of gas. Early the next morning he began a hike to Austin for two gallons of gasoline, arriving back at the Little Giant at 3:30 in the afternoon. By monitoring our odometer we could estimate the spot where he sputtered to a stop. We also noted that ten of those miles were up the mountain and ten back down. Remember, this is the "Loneliest Road in America" It was even more so in 1915.

While Thomas Jefferson and his buddies in the Continental Congress were blazing new trails with the Declaration of Independence in Philadelphia in 1776, Fray Francisco Garce's and another monk became the first white men to enter what would later become Nevada, setting out to create a new trail from the Colorado River to California in 1769, and Monterey in 1770. Their efforts were hindered by the hostilities of Indians, rugged terrain, and severe weather. The next white men into the area were the rugged mountain men spurred on not by gold or silver but by furs to satisfy European fashion demands for fur hats. These men learned how to survive in the wilderness, made friends with the Indians, and became a credible source of guides for wagon trains and explorers.

The first American in this category was Jedediah Strong Smith, a hardy, rugged experienced man of the wilderness—a true Mountain Man. At one point while leading a party he was attacked by a grizzly bear which ripped off his scalp from above his left eye across the top of his head and tore off his right ear—he also received several broken ribs. But he was a tough man and in spite of the pain was able to instruct Jim Clyman how to stitch his scalp and ear back in place. Smith and his friends later went into the freight business, hauling supplies to the Southwest. They once ran into a particularly bad dry area and Smith set out to find water. Some say he was killed by Commanche Indians. Jedediah Smith was remembered, not only for being the first American to enter Nevada, but for having learned first-hand more about the West than any other person of his time. He was also the first man to make known any information about the great desert wilderness of what today is Utah and Nevada.

In 1861 President Buchanan signed a bill which created the Territory of Nevada. Two days later, Abraham Lincoln was sworn in as the sixteenth President of the United States. On October 31, 1864, Lincoln signed the bill to make Nevada a state.

Loneliest Road

"The Loneliest Road in America," 20 miles east of Eureka, Nevada.

EUREKA, *Nevada*

IN HIS OWN WORDS:

Friday, October 29, 1915

Traveled 30 miles East of Eureka. Made 108 miles. Had bad going over chuck holes in flats.

1864 EUREKA 1964

EUREKA! A MINER IS SAID TO HAVE EXCLAIMED IN SEPTEMBER, 1864 WHEN THE DISCOVERY OF RICH ORE WAS MADE HERE—AND THUS THE TOWN WAS NAMED. EUREKA SOON DEVELOPED THE FIRST IMPORTANT LEAD-SILVER DEPOSITS IN THE NATION AND DURING THE FURIOUS BOOM OF THE 80'S HAD SIXTEEN SMELTERS. OVER 100 SALOONS. A POPULATION OF 10,000 AND A RAILROAD—THE COLORFUL EUREKA AND PALISADE—THAT CONNECTED WITH THE MAIN LINE NINETY MILES NORTH OF HERE.

PRODUCTION BEGAN TO FALL OFF IN 1883 AND BY 1891 THE SMELTERS CLOSED. THEIR SITES MARKED BY THE HUGE SLAG PILES AT BOTH ENDS OF MAIN STREET.

Nevada centennial marker no. 11

THE 1964 CENTENNIAL MARKER nearly says it all. But not quite! Eureka is one of the Lincoln Highway towns not indebted to a railroad for its existence. The invention of Eureka began with America's first important discovery of silver-lead, in 1864. The town boomed in 1869 when the Eureka Consolidated refinery patented a successful method of separating silver and lead. By 1878 Eureka boasted a population of more than 9,000, beginning like many mining towns as a city of tents. By the mid-1860s there were 250 buildings and a post office, always a symbol of prosperity. With five years of growth and prosperity, Eureka was claiming to be the second largest urban center in Nevada. There were dozens of saloons, gambling houses and bawdy houses, three opera houses, two breweries, five volunteer firefighting companies as well as the usual complement of doctors, bankers, lawyers, merchants, hotels, newspapers and other businesses. There were 50 mines producing lead, silver, gold and zinc. The smelters were capable of producing more than 700 tons of ore per day as they belched black smoke into the air, poisoning local residents and desert vegetation. It was no accident that Eureka was called the "Pittsburgh of the West."

Between 1876 and 1880 Eureka suffered four major fires causing millions in damages, rebounding after each and rebuilding thick-walled brick buildings with iron fire doors and installing hydrant systems throughout the town. The natives responded to the fires with tongue in cheek, accepting them as opportunities to build bigger, better and fancier buildings, which in fact benefited succeeding generations. Many of these buildings today are listed in the National Register of Historic Places. An example is the 1879 Eureka County courthouse, which still houses

Eureka County Courthouse

Eureka County Courthouse. Construction began in 1879 and completed in 1880 at a cost of approximately $55,000 plus $15,000 for the adjoining jail. It was the finest courthouse in Nevada, outside of Virginia City. It is still used today and visitors can still marvel at the judge's Spanish cedar bench, tin ceiling and chandeliers

the original courtroom, complete with original wainscoating, pressed tin ceiling and chandeliers—and a Lincoln Highway concrete post in the sidewalk at the front of the building.

The homogeneity of American population evidences itself again in Eureka with Irish, German, Italian, Chinese, and Jewish immigrants. Their widely divergent vocational choices contributed greatly to the economy of the community. The Italians became known for the manufacture of charcoal which was used in the many smelters in town. This production of charcoal eventually consumed every tree within 50 miles. The Chinese tended to work in food services, laundries, medical practices, and as laborers. Nearly sixty percent of the population in Eureka was foreign born.

Courthouse

Lincoln Highway post in front of the Courthouse. Note the bell, one of two, which were rung by fire volunteers as fire alarms.

Eureka, stuck in the waist of Nevada, was not unmindful of the world beyond the eight mountain passes of Nevada and its wide-open deserts. Eureka was turned on by the imminence of a coast-to-coast highway coming to town, which was granted considerable "footage" in the *Eureka Sentinel* for the promotion of the Lincoln Highway. The following are three excerpts from 1915, the year my father came to town;

THE EUREKA SENTINEL

May 22, 1915

Lincoln Highway Travel

The travel over the Lincoln Highway has evidently commenced in earnest for a number of cars have passed through Eureka this week. While several have gone through from the east to the coast, most of them are eastern cars homeward bound having taken the southern route to the expositions and, after touring California are making the return trip over the Lincoln Highway. Two families returning East in Cadillac and Ford cars stopped at the Brown Hotel last night and contiued their journey this morning.

Eureka Sentinel

Eureka Sentinel Museum, the office of the Eureka Sentinel *Newspaper, was built in 1879 to replace the former office, which was destroyed by one of Eureka's major fires. The first issue came off the press in 1870, with the* Sentinel *continuing publication for ninety years, until 1960.*

August 28, 1915;

Russell Advocates The Lincoln Highway

Total Abstince Lecturer Hopes to see Congress Aid Project. Preaching total abstinence, with a mention of good roads on the side, the Rev. Howard H. Russell, founder and one of the superintendents of the Anti-Saloon League of America, arrived here Thursday as previously announced on his "Water Wagon tour" of the Lincoln Highway, from New York to San Francisco."

"I'm certainly enthusiastic about the Lincoln Highway and as a citizen I will do everything I can to induce Congress to make a Federal appropriation for the road." Said Dr. Russell. "I have come all the way from New York on the Lincoln Highway and have not missed an engagement."

September 18, 1915

TO AID LINCOLN HIGHWAY

Eastern women to give Lecture in Eureka Tuesday Evening, September 21.

The SENTINEL has been requested to announce that Mrs. Lee C. Boardman and Mrs. Sarah McDonald of the executive committee of the Lincoln Highway Women's Auxiliary will give an illustrated lecture in Eureka Tuesday evening, Sept. 21, of scenes from New York to San Francisco over the Lincoln Highway.

Consul Ralph Zadow of the Lincoln Highway is authorized to announce that a small admission sufficient to cover the expenses of these ladies while stopping over in Eureka, will be charged.

Efforts will be made to interest the women of Eureka in the Auxiliary designed to promote interest in the complete establishment of the Lincoln Highway from coast to coast.

Modern Eureka is at 6,500 feet elevation and the Lincoln Highway still goes smack dab through the center of town. Eureka is the county seat for Eureka County and contains the majority of the county's 1,900 residents. If you like fresh high desert air, stunning mountain views, no traffic congestion and some of the best-preserved history of United States mining, you'll love Eureka. And Eureka will love to have you.

Press Room

Press room of the Eureka Sentinel. *Most of this is the original equipment and presses. Many of the posters on the wall date back to the 1880s.*

First License

First auto license plate in Eureka County in 1913. It was found in a dump in 1963.

AUSTIN, *Nevada*

IN HIS OWN WORDS:

Saturday, October 15, 1915

Fair. Made 118 miles. 14 miles from Austin ran out of gas 10:00 P.M. Had rough desert roads. Lots of chuck holes.

Sunday, October 17, 1915

Fair. Got up early. Left camp on foot to Austin for 2 gal gas. Over rough mountains 14 miles. Total 28. Got back to camp 3:30 P.M. Run on past Austin 15 miles. Total 29 miles.

[on his way back East]

Thursday, October 28, 1915

Fair. Traveled from French Men's Station to 10 miles East of Austin. Bad holes over flats. 84 miles. Got lost in desert. Got wrong trail.

Friday, October 29, 1915

Left camp E. of Austin at water trough.

---

*L*ike an Easter egg hidden on a billiard table, Austin is hard not to find. All motorists traversing U.S. Highway 50 eventually funnel onto Main Street, Austin, whether they want to or not.

Most travelers, by then either hungry or out of gas, will welcome it as a place to fill up and stretch their legs.

Others will regard it as a bottleneck forcing them to slow down and pay attention for a moment.

A very few will quit their jobs, buy a house, and stay in Austin for the rest of their lives.

THAT IS THE WAY JIM ANDERSEN in *Lost in Austin* introduces Austin, Nevada. The importance of Austin depends on who you talk to. Descriptions vary from "One horse town"—it was discovered by a horse—to "semi-ghost town," which means the two filling stations are run by live people, to "God's country,"—who else? Legend has it, backed up by some incredible history, that in 1862 a horse belonging to a William Talcott, kicked up a piece of quartz containing silver and gold. The details here are a bit murky but let's take their word for it. We do know that mining became a big deal to Austin (named for Austin, Texas), collecting a population of 7,500 one year later. To keep these people occupied, a lumber mill was built, along with 400 homes, schools, churches, hotels, stores, and the required number of saloons. The nemesis for most pioneer Lincoln

Gary Elam.

Austin, Nevada, *sparkles after a refreshing summer rain*

Highway towns was fire, but not so Austin. The nemesis here was flood. Cloudbursts, mostly unannounced, created flash floods which rolled huge boulders down Main Street, smashing freight wagons and anything else in their paths. Second-story stairways were hinged to swing upward, out of the way.

By 1880 the mines were showing signs of fatigue and soon Austin's total $50 million in ore production was history. Austin's isolation has helped preserve it today as an excellent example of things as they were.

The first designated highway across the state was not, sad to say, the Lincoln Highway even though it happened in 1913. It was the northern competitor to the Lincoln Highway from Wendover to Verdi. It not only horned in on the Lincoln Highway's prestige but tried to capitalize on the Lincoln name as well. It was called, "Lincoln Trail." The following excerpt lets us know how the locals felt about it. From the August 14, 1915

REESE RIVER REVEILLE

Autoists Switch over to the Lincoln Highway

Reluctantly, the Reveille again must speak of the short-sighted talk of people along the Lincoln Trail regarding conditions on the Lincoln Highway.

A party switched from Battle mountain to Austin Monday to get out of the Lincoln Trail on to the Lincoln Highway. They reported that there were long stretches along the Humbolt River where they could make only from two to four miles an hour. When they hit the gravel roads of the mountains, on the Lincoln Highway, they could make twenty or thirty miles an hour. That is their testimony.

Wednesday, G. E. Bowerman, a businessman of Anthony Falls, Idaho, with his family, went east through Austin after seeing the Exposition. Going west, he traveled the Lincoln Trail. He said that in Wells, Elko, Deeth, Winnemucca and other towns along the way he was told that gasoline in towns on the Lincoln Highway was fifty cents per gallon; that the dealers overtanked them (what a joke), and that the Lincoln Highway people were hold-ups from end to end. "We have had delightful going on the Lincoln Highway," Bowerman said.

Ironically the northern route ultimately became the route of choice, becoming Interstate I-80 in 1958.

The superiority of the Lincoln Highway, at least in the minds of Austin-ites, was evidenced by the Lincoln Highway Moving Picture Caravan showing up in town on Monday, August 16, 1915. The following report appeared in the August 21, 1915 issue of the

REESE RIVER REVIELLE

Taking Movies Of Lincoln Highway

Monday afternoon the Lincoln Highway Moving Picture Caravan, consisting of five cars and fourteen people, reached Austin. The party left New York City on May 15 and expect to reach the Pacific Coast on the 25th of this month.

The purpose and object of this trip is to take moving pictures of towns and cities, scenery, road conditions and other interesting points which are encountered by transcontinental tourists on a trip from one coast to the other and in this manner to bring before the eyes of the American people the wonders and beauties which only can be realized by one who has covered the ground.

Gridley Museum

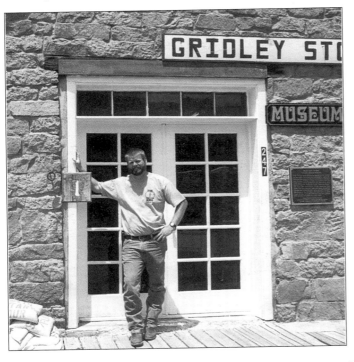

Rev. Ron Barney stands in front of the historic Gridley building in Austin. In 1863 it was a general merchandise store operated by the firm Gridley, Hobart and Jacobs. Gridley is best remembered for his 1864 wager which prompted the auctioning of a sack of flour for donations to the "Sanitary Fund," the Civil War forerunner of the American Red Cross. The flour was sold again and again throughout Nevada and California, then taken east and eventually auctioned at the St. Louis Sanitary Fair in 1864, in all raising about $275,000. The building is now the Gridley Store Museum, opened just one week before our visit June 23, 2001. Note the symbolic sacks of flour at Rev. Barney's feet.

Clouds cover the top of the mountain which harbors Austin, Nevada. This is the approximate spot where my father ran out of gas fourteen miles from Austin.

Upon arriving here, H. C. Ostermann, consul at large of the association, and head of the expedition stated:

"The condition of the entire Lincoln Highway can be said to be good. In fact many tourists who have been encountered along the way have remarked about the splendid roads which they have traversed, especially in this western country where the uninitiated traveler expects to find the worst conditions. The greatest surprise of the whole trip for the Lincoln Highway has been the road from Salt Lake City to Austin."

Since 1913 was a "Whoopie!" year for the Lincoln Highway, it should be noted that this was the year the State of Nevada passed its first motor vehicle law. The law stated that the licence tag fee would be 12.5 cents per horse power, minimum of 20 horse power.

My father was also in Austin in 1915. The first time was October 16, 1915, when he ran out of gas and had to walk 14 miles to Austin for two gallons of gas, and again on October 28, on his return trip. This time he made a wrong turn east of Austin and got lost in the desert.

You can get lost in Austin but you will have to work at it. The altitude is 6,575 feet, the same as when the population was 7,500, but little else is the same. The approximately 300 residents are scattered among ranching, mining, self-employment and government jobs with the economy mainly tourist based. The town has one hardware and lumber store, two gas stations, one auto repair, two restaurants, three motels, (which says something about the transitory nature of things), one bed and fix-your-own-breakfast, one trailer and RV park, and two laundromats. There are three bars (down a tad from 1863), two antique shops, three rock, gem and gift shops, two real estate offices, and one mountain bike shop. All of Austin's businesses are family owned and operated. No Wal-Mart, K-Mart, or Target.

International Hotel

International Hotel, one of the oldest buildings in Nevada, was torn down piece by piece and moved from Virginia City to Austin in 1862. Famous for its old ballroom and dining room, it was the center of many wild discussions of the riches and failures and the exciting events of the times.

Loneliest Road

Lincoln Highway going west from Austin on the "Loneliest Road."

FALLON, Nevada

IN HIS OWN WORDS:

Monday, October 18, 1915

*Warm. Traveled on to 14 miles W. of Fallon.
Made 119 miles. Long sandy stretch.
Got stuck in sand.*

Tuesday, October 19, 1915

*Clear, cold. Left camp 14 miles west of Fallon.
Had broken spring fixed it up and traveled to beyond
Truckee, Cal. 95 miles.
Good mountain roads.*

ONE HUNDRED AND FIFTY YEARS AGO Paiute Indians hunted pine nuts in the mountains and chased deer over the same ground where today people pursue adventure, business, and vacations on the Lincoln Highway. While the Lincoln was still wet behind the ears, traders were already chasing their dreams, being captive to an insatiable curiosity. These happy valley hunters complained about dust, chapped lips, desolation, cold, heat, and the shortage of amenities associated with civilized living in the East. The Paiute, if by some means still to be discovered, could encounter the Lincoln Highway, they would no doubt complain about burning eyes from horseless buggy exhaust, noise, terrorized game, and loss of freedom.

Following the Paiute, the hunters of furs crossed this land. Then came the seekers of gold who in 1849 profoundly affected the land which we now call western Nevada. The discovery of silver in 1859 put this area on the map.

At the turn of the century there was no Fallon. There was, however, the Mike Fallon Ranch, settled by Mike and Eliza Brunner Fallon in 1896. The postoffice was in their home. Following President Theodore Roosevelt's Land Reclamation Act of 1902, Mike Fallon sold his ranch to Senator Warren W. Williams who had the land plotted and began advertising the sale of lots in 1903. Thus began Fallon's first housing development (there goes the neighborhood) which was greatly enhanced when the Senator pushed through legislation to have the county seat moved from Stillwater to the new town of Fallon.

The growth and prosperity of Fallon were assured with the construction of the Derby Diversion Dam, on the Truckee River, and the Lahontan Dam. The Derby Dam was formally dedicated on June 17, 1905, and the city of Fallon was incorporated in 1908, with the I. H. Kent Mercantile Company as one of the earliest business establishments. This name catches the attention of Lincoln Highway officials because I. H. Kent was appointed by the then state consul Hoag to represent Lincoln Highway interests in Churchill County. The

Churchill County Museum and Archives

Railroad Construction

Volunteer railroad construction crew on Maine Street, July 23, 1912. The crew is on its way to begin grading for the Fallon Electric Railroad line—a project never completed.

Main Street

Main Street, Fallon, Nevada in 1915. The Barrel-House was one of the thirty-four contributors who donated toward the expenses of putting Fallon in the Lincoln Highway film. The donation totaled $99.50. Surely someone added an extra fifty cents.

prestige of the Lincoln in the county and in Fallon was largely the product of the dedication and contagious enthusiasm of Kent.

Fallon has flown its flag over many successful enterprises over the years, not the least of which was the "Hearts Of Gold" cantaloupes. In fact Lahontan Valley was known for "The King of Cantaloupes," large-sized melons with unmatched flavor. The drought years of 1931 and 1932 drove farmers back to raising high protein alfalfa, their most stable source of farm income.

Modern Fallon has a population of 8,200, is 287 miles from San Francisco and 3,100 miles from New York City.

My father mentioned Fallon on his way west on October 18 and 19, 1915. The weather he reported as warm on the 18th, but clear and cold on the 19th. He also experienced a broken spring, which he fixed (he didn't say how), and then drove on to Truckee, California. This experience springs to mind (pardon the pun) a similar one in the same area by another traveling couple a year earlier in 1914—probably not caused by the same chuck hole, but who knows? Their broken left rear spring carried

Overland Hotel *(1908) on the corner of Nevada Street and East Center Street. The Lincoln Highway filming crew stayed here while shooting up the town with their cameras in August 1915.*

Overland Hotel, *2001. After 98 years, the Overland has changed little. Note the Lincoln Highway sign on the front pillar.*

Leon Schegg

Five Steeds

The five steeds of the famous "Three-Mile Picture Show" line up just outside Omaha, Nebraska. l to r: Stutz, Studebaker, Packard, Oakland, Little Giant. None has survived to the present. Packard and Studebaker merged for a brief time (1952 – 1956), Stutz threw in the towel in 1935, and Oakland morphed into Pontiac in 1932. The Little Giant yielded to the manufacture of riveting equipment in 1918.

much more drama than my father's. He used only six words to record it. For the 1914 couple it was an opportunity to vent their displeasure about the hotel in Austin, presumably the famous International Hotel, and the owner whom they described as a "horrid old Mick" who also happened to be Senator Easton. The Senator's son was a "horrid red-head know-it-all." The mechanic across the street looked like a "cut-throat," so they shook the dust from their feet and drove 70 more miles to Eureka for repairs.

In those days anything on the Lincoln Highway became grist for the *Churchill County Eagle*:

THE LINCOLN HIGHWAY

Saturday, June 19, 1915

President of Lincoln Highway Association visits Fallon

Fallon was visited Tuesday by the President of the Lincoln Highway Association, who is also president of the Packard Auto Co., and party who were making a tour from Detroit to the coast in a big new 12-cylinder Packard, known as the "Twin Six." The party stopped at the overland hotel for luncheon, and several Fallon people called to pay their respects and to view the big car and its new engine.

Distinguished visitors were Henry B. Joy, president; A. F. Bement, manager of the publicity department of the Lincoln Highway Association, and E. Eisenhut, driver.

Saturday, July 24, 1915

Little Boy Breaks Arm On The Lincoln Highway

A little boy, riding on the running board of an auto going eastward over the Lincoln Highway, fell off a few miles west of town Sunday evening and broke his arm. The boy and his parents were traveling with a party of three families in three autos. The little fellow who was about nine years old, was brought to town and the arm set by Dr. Ferrell. The people stayed over night at the Golden. The little fellow cried all night with pain. The party resumed their journey the next morning for their home in Salt Lake City.

Saturday, August 28, 1915

John Bruns Follows Lincoln Highway From New York To San Francisco

To make the trip from New York to San Francisco on a bicycle is no easy task. At least that is what John Bruns of Port Jervis, N.Y., says. He left New York City May 3 on an ordinary bicycle and following the Lincoln Highway had covered 3,350 miles when he reached Fallon. He says the Lincoln Highway is a wonderful project and he doesn't see how anyone could get through the country without it.

In the East he was delayed a great deal by rains, hav-

Churchill County Museum and Archives

Frenchman Station

Frenchman Station (circa early 1900s) was about thirty miles east of Fallon on the Lincoln Highway. It is frequently mentioned in notes and diaries of many early Lincoln Highway travelers. According to the diary of Paula Davis in 1914, Frenchman Station was so named for the owner, a small Frenchman Aime Bermond. There is no local water source so Bermond maintained a deep cistern which he supplied with water twice a week from twenty-five miles up in the mountains. He charged twenty-five cents a bucket for it. My father camped two miles west of here on October 27, 1915. He records, "Bad sandy stretches. Stuck in salt flat."

ing to stop some times from four to five days in a place. When he left New York he had two companions who rode with him over the first 800 miles to Indianapolis. Since then he has been alone. "Believe me," he said, "I have enjoyed the hospitality of the West." He does not doubt that he will remain west of the Rocky Mountains, being attracted by the whole-souled people of the West.

He expects to get to San Francisco by September 3, making a four-month trip with a total distance of about 3,700 miles. He carries baggage to the amount of 93 pounds on his wheel and rode 3,100 miles without a puncture.

Saturday, October 16, 1915

Lincoln Highway Affairs

H. C. Ostermann, Counsul at Large for the Lincoln Highway, accompanied by Mrs. Ostermann and his private secretary, stopped in town Tuesday night and held a conference with County Commissioners, Ernst and Harrigan, Mr. Benadum being absent. Mr. Ostermann had charge of the party that passed through here last August taking moving pictures along the Lincoln Highway, and states that all of the films taken in Fallon and on the Truckee-Carson project were excellent.

Father chugged into Fallon just two days later on October 18 and camped fourteen miles west of town where he got stuck in sand.

Saturday, December 18, 1915

Had Queer Auto Luck

Whiskey For Gasoline and Instead of Honk, Honk—It Was Hic Hic.

C. W. Kinney, accompanied by R. W. Wiley, had

a peculiar experience coming in from Eastgate a few days ago. They were short of gasoline. The gas was too low to reach the carburetor and, after various schemes to get help, Mr. Kinney happened to remember that they had secured a bottle of whiskey at Frenchy's. He reasoned that it could do no harm to empty the bottle into the \gasoline tank, so in it went—except a small amount allowance to be used in generating the engine De Chauffeur. The next thing was to crank the machine and before a half turn had been made, away the engine went, and they came flying into Fallon, leaving an odor of corn juice behind them. The machine wobbled some, and when the warner was sounded, instead of it going Honk, Honk, it went Hic, Hic.

There were no Oscars in 1915—or Emmys either—only a few heady Lincoln Highway entrepreneurs who spent their daytime hours and nighttime dreams on their newest obsession. It was Arthur Pardington, the association secretary, believe it or not, who came up with the latest far-out scheme—a stunt to do Carl Fisher proud: make a movie of the Lincoln Highway. That's right, all 3,384 miles of it from New York to San Fransisco! Of course Henry Joy and Fisher signed on. Henry C. Ostermann, field secretary, became point man for the adventure, having traveled the highway several times before and knew the local movers and shakers, proving the old adage, "It's who you know, not what you know."

There are always opportunists eager to cash in on something new and with such a marketing expert as Fisher twisting their arms, who could refuse? The Stutz Company responded with a lead car for the Ostermanns, his secretary and official cinematographer. Studebaker showed up with a nice machine for the research engineer and publicity department. Next, of course, came the luxury division with—what else,—a Packard, for dignitaries like governors, mayors and such. Oakland got in the act with a car in the Midwest to help transport the overflow (the idea was growing like a snowball in January). Then last, but not least, a Little Giant truck from Pneumatic Tool joined the entourage in Omaha to lend a hand with logistics. The Little Giant came equipped with drivers, Leroy Beardalay (or Beardsley) and Earl Phillips (or Philips), provided by the Chicago Pneumatic Tool Company. Only two of these honored steeds (Studebaker and Little Giant) were manufactured on the Lincoln Highway—Studebaker in South Bend, Indiana and Little Giant in Chicago Heights, Illinois.

Advance notices were sent so each town could have its own fifteen minutes of fame with its best foot forward. Fallon, Nevada, "A remote Nevada town", ran with it like a kid with a kite. In fact on Christmas day in 1915 the following headline appeared in the

CHURCHILL COUNTY EAGLE

Lincoln Highway Motion Pictures

Over 16,000 Feet of Film to be Exhibited in Fallon Soon, Will Run Both Day and Night to Give Everyone a Chance to Traverse the 3,384 Miles From New York to San Francisco.

The caravan did show up in San Francisco and the three-mile picture show earned its keep as a significant part of the Lincoln Highway's exhibition in the Palace of Transportation. The film was shown continually until the Exposition closed. The caravan stopped again in Fallon on its return trip, again patronizing the Overland Hotel while the locals of Fallon and Churchill County dropped everything, even schools closed, to admire themselves on the screen at the Rex Theater on January 3, 1916.

Churchhill County Museum

Concrete post at Churchill County Museum, 1050 Maine Street, Fallon, Nevada. The Lincoln Highway sign is from the California State Automobile Association..

My Auto, Without Thee, In Money I Would Be

My auto, 'tis of thee,
Short cut to poverty –
Of thee I chant.
I blew a pile of dough
On you three years ago,
Now you refuse to go –
Or won't or can't.

Through town and countryside
I drove thee full of pride;
No charm you lacked.
I loved your gaudy hue,
Your tires round and new –
Now I feel mighty blue,
The way you act.

To thee, old rattlebox,
Came many bumps and knocks;
For thee I grieve.
Badly thy top is torn;
Frayed are thy seats and worn;
The croup affects thy horn,
I do believe.

Thy perfume swells the breeze,
While good folks cough and sneeze,
As we pass by.
I paid for thee a price,
'Twould buy a mansion twice;
Now everyone yells "Ice" –
I wonder why.

Thy motor has the grip;
Thy spark plugs has the pip,
And woe is thine.
I, too, have suffered chills,
Fatigue and kindred ills,
Trying to pay the bills
Since thou were mine.

Gone is my bankroll now;
No more 'twould choke a cow,
As once before.
Yet if I had the yen,
So help me John – Amen!
I'd buy a car again
And speed some more.

Bay City Motorist • From EARLY AMERICAN AUTOMOBILES, circa 1915

CALIFORNIA
And The Lincoln Highway

29 Towns, 1915 137 miles, 1924

"CALIFORNIA, HERE I COME" HAS DRUMMED the ears of countless adventurers from time before we may have any interest. The drumbeat has always emanated from the West to restless souls in the East, coming from as far as Europe and beyond. California was as far west as they could go without a ship, but some, in fact, did go by ship—the Mormons in 1846. This idea of westward movement is not the result of any scientific research. It comes only from my imperfect observations. The exceptions of course are the adventurous Indians who emigrated east from Asia across the narrow straits into Alaska.

California has bragging rights to great diversity both human and geographic. For purposes here, credit will be limited to whatever has a good word for our favorite link—Lincoln Highway. If you're into rocks, California can give you rocks. On the Lincoln Highway you cannot go to Mt. Whitney (14,494 feet) because the Lincoln Highway has better things to do, but you can drive through the Sierra Nevada. In fact you had better pack your history bag because Donner Pass was the cause of not a few tragic human dramas. The most notable was the Donner Party in 1846-47, but let's not give the Donner tragedy credit before credit is due.

Two years before the Donner tragedy, in 1844-45, the Stephens-Murphy-Townsend Party was the first to successfully challenge the snow, cold, rocks, and cliffs of what became the Donner Pass. The drama began with a blacksmith from Georgia in an Indian subagency in Council Bluffs, Iowa, who cocked his ears to the sound of Murphy and Townsend wagons entering town. Since Stephens had fur-trapping experience in the Northwest, the magic words, "California" from the lips of Murphy and Townsend, stoked a smoldering fire in the blacksmith

Donner Lake

Donner Lake as seen from Interstate 80.

Snowsheds and Lake

Railroad snowsheds as they appeared in 2001. Note the replaced top on tunnel #7. To the left of #7 is the Lincoln Highway remnant my father traveled October 20 and 26 in 1915

and he became the "Stephens" in the Stephens-Murphy-Townsend Party. His knowledge of the plains and mountains earned him the position of captain of the entire caravan numbering twenty-eight men, eight women and fifteen children.

Like honey bees to honey, the enthusiasm and visions of California as a potential for wealth and happiness attracted even Shoshone and Paiute Indians. When one old Paiute kept repeating the word "Tro-kay," the immigrants assumed he was saying his name. What he was saying, it turned out, was "Everything is all right," because of his concern they might think he was hostile. "Tro-kay" then became Truckee, and he it was who led them to a river leading into the Sierra—a fortuitous choice, which saved the party precious time as well as lives, and did not go unnoticed. In gratitude they named the river Truckee. Even so, the Sierra Nevada were a severe challenge to the courage and ingenuity of the party. When near the top, six wagons were left behind along, with three men to guard them until they could be reclaimed in the spring. The other five

Central Pacific Railroad Photographic History Museum, @ 2001 CPRR.org.

Donner Lake

The A.J. Russell stereoview of the Central Pacific Railroad tunnels and Donner Lake in 1868. Andrew J. Russell had been an official Army photographer during the Civil War and was commissioned by Union Pacific to photograph its construction process. Note the original cap on tunnel #7 (center). The top of the tunnel was later removed but then restored because heavy snow filled the gap.

wagons were unloaded and hauled up the cliffs with chains, and were the first to cross the Sierra Nevada into California. The trio left behind for guard duty were provided a "half-starved withered cow" and hastily built a cabin one mile south of Donner Lake, twelve by fourteen and eight feet high, made from pine saplings with a brush and rawhide roof, one window, one door, and a crude chimney, all constructed in two days. At the completion of the cabin and in spite of heavy snowfall the day following, the trio had a change of heart and decided to take out after the main party. Moses Schellenberger, however, soon became exhausted and returned to the cabin where he survived the winter on ox hides and trapped foxes, but survive he did until late February 1845. Then he was retrieved and safely entered the promised land. There were no human casualties. In fact the party had grown by one—Ellen Independence Miller, born at Independence Rock, Wyoming.

Two years later, in 1846 and 47, the Donner Party became the most famous to cross the Sierra Nevada. In April 1846 two Illinois farmers, George and Jacob Donner, could not resist the draw of the West, and even though they were in their 60s they packed their families into six wagons and headed for California. Their entourage included James Reed, a well-to-do cabinet-maker and close friend, along with his family. The group soon numbered ninety-one people with twenty wagons headed for utopia, but stumbled instead into a cold, white abyss. Their first mistake was to take a recommended shortcut which promised to save them 350 miles (famous last words). The "shortcut" cost them invaluable time, oxen, cattle, wagons and most important of all, morale. They entered the Sierra Nevada fractured in spirit and ill prepared for normal mountain circumstances and certainly not for an early winter storm requiring survival skills. Only forty-nine of the original ninety-one ever saw the fruitful fields of their dreams—California. Those who survived did so only after unimaginable suffering of body, soul, and

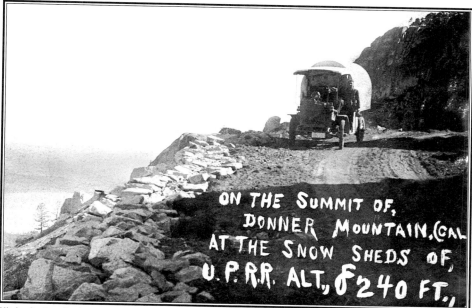

Nissley family collection

Little Giant

My father stands on the running board of his Little Giant just around the bend from the Central Pacific Railroad snow sheds. Note Donner lake in left bottom corner.

Courtesy of Kathy Franzwa

Lincoln Highway Fans

Lincoln Highway fans, from the tenth annual LH conference in Sacramento, California, hike down the LH remnant, where my father took a photo of his truck October 26, 1915.

Rainbow Bridge

The Rainbow Summit Bridge on Donner Pass Road, also U.S. 40.

mind. One survivor later died in California, a tormented half-crazed man because of intolerable memories and feelings of guilt. The survivors were ultimately rescued by four separate rescue parties, beginning in March of 1847.

If sometime you are privileged to enjoy the amazing beauty of Donner Lake from high up in the pass, pause, take a deep breath, close your eyes and try to imagine the stories the lake could tell of freezing limbs, loneliness of heart, anger, despair, guilt, starvation, and even cannibalism. Some say the lake should have been named for the Stephen-Murphy-Townsend Party as a memorial to success, rather than to one of defeat and tragedy. On the other hand, is not the lake a fitting memorial to those brave people who joined the thousands before them seeking their own Holy Grail? In any case the Lincoln Highway in all thirteen states treads on millennia of human drama and memories, most of which are long forgotten but nevertheless provide a common bond linking us together and fashioning us into who we are.

Pitts Joyner

Dean "Pitts" Joyner from the Sacramento Trail Society stands by his trail bike. We met up in the rocks of Donner Summit where he was monitor for a mountain bike race soon to begin. Note the markers on the rocks where the bikers must perform and be judged by their skill. Mr. Joyner said bikers come from all over the world to accept the challenge of Donner Pass.

What else do you want to know about California? Having negotiated the Sierra Nevada and Donner Pass and now coasting into the land of milk and honey—well, vegetables, peaches and such—what now would you like to know? Probably not a reminder of the San Andreas Fault and its like, which may someday shove a huge chunk of the state into the Pacific.

On a more cheerful note, the wonderful climate of California and other coddling qualities were enjoyed by more than 100 different Indian tribes thousands of years before the Spaniards arrived. The Spanish tried to colonize California from the south as early as 1533. Their eventual opulent trade success attracted the English and French paving the way for the eventual Lincoln Highway. In 1848 the new landlords (United States) took charge with the Treaty of Guadalupe Hidalgo.

The discovery of gold in 1848 helped pave the way for the Lincoln Highway 65 years later, by infecting the East with "California Fever," motivating anyone so inclined to head for the cure in California. The completion of the transcontinental railroad in 1869 ended the state's isolation. By 1890 San Francisco was the largest city on the Pacific Coast and a commercial center on a world scale. By 1915, the year my father drove the Lincoln to San Francisco, California promos sounded much like modern hype. "Land of sunshine and flowers" and "The playground of America" are blurbs in the 1916 *Complete Highway Official Road Guide of the Lincoln Highway.*

Modern California is populated by 33,871,648 inhabitants, with 217.2 people per square mile. Compare this with 5.1 for Wyoming. In 1915, California could boast of only 3,300,000 inhabitants. It became the thirty-first state September 9, 1850, as a direct consequence of the discovery of gold in 1848. It is now the third largest state, encompassing 163,707 square miles with 840 miles of Pacific coastline. Agriculture and related activity today contribute only three percent to the state's economy yielding to manufacturing and commercial's seventy-eight percent. This in part explains the transition from a rural population to the present ninety-three percent who live in areas defined as urban. The state experimented with several locations for a capital before landing on Sacramento in 1854.

Lincoln Highway Garage

Lincoln Highway Garage, Portola Avenue and North, Street in Livermore, California.

The 1924 edition of *The Complete Official Road Guide of the Lincoln Highway* says it well:

Climate, that will-o'-the-wisp that has led travelers in search of comfort or health a weary chase over many lands, comes closest to the ideal in California, where each individual can select that which suits him best.

Auburn City Hall

One of the original 1928 concrete markers (not in the original location) stands in front of the Auburn, California City Hall.

LOOMIS, *California*

IN HIS OWN WORDS:

Wednesday, October 20, 1915

Traveled on to Loomis 87 miles. Was coasting practically all from Donner Lake west.

Thursday, October 21, 1915

Fair. Left Loomis and camped at fruit farms.

IMAGINE, IF YOU CAN, THE FEELINGS of those survivors of Donner Pass who so recently endured the cold, fought huge rocks, cliffs, and mountains of snow, and now before them lie vistas of clear sky, warm breezes and fruit farms as far as the eye could see.

It caught my father's attention on October 20, 1915, enough to record: *"Left Loomis and camped at fruit farms."* His Lancaster County, Pennsylvania, farmboy genes rose to the surface and fruit farms looked pretty good. He did not say what fruit he ate, but eat he did. The quality and flavor of California fruit also caught the attention of fruit lovers in the East and throughout the world. Placer County became the birthplace of the fruit industry in California. Decades later I, too, savored the gustatory California nectarines, peaches, et. al. even to the discomfort (to put it delicately) of too much of a good thing. Virtually everyone who grew up in Loomis spent summers working in the fruit packing sheds.

But not every immigrant ate fruit. Griffith Griffith from Wales arrived in California in 1853 and survived on granite. In 1864 he leased land from the Central Pacific Railroad for a granite quarry, and in 1865 founded the town of Penryn. He came to California not for fruit but for two economic reasons: the existence of high quality granite and close proximity to the CPRR. A spur track to his quarry was completed in 1870—oxen pulled cars up the hill and gravity took them down.

Granite Museum

Granite Museum, former office for the G. Griffith Granite Works, which was established in 1864.

Shirley Patocka

Shirley Patocka, volunteer, stands in front of the G. Griffith Quarry Museum.

In 1867 Griffith married Julia Logan, a widow with four children. They had no children of their own, so at his death his widow sold the quarry to David Griffith, a nephew, who trekked back to Wales to marry his childhood sweetheart, Sydney Davies, and brought her to California. They had one child, Enid, who never married. All this family detail is to explain how Placer County got its hands on the property. In 1976, Enid died at the age of 81, with no heirs, and left the quarry property to the people of Placer County. This is important because the quarry is on the Lincoln Highway, a fact anyone could guess.

The quarry ceased operations at David's death in 1918, when he was 56. The Placer County Parks Department opened the grounds to the public after creating trails and building fences around the quarry holes. The granite office building was restored and opened to the public on May 8, 1981, with artifacts from Enid's estate and Friends Of The Quarry on display. Granite from this quarry has held up important buildings far and wide, notably the San Francisco Mint, constructed in 1877. As proof of its mettle, it rode out the 1906 earthquake with no damage.

Fruit Growers

Loomis Fruit Growers as my father would have seen them.

Depot

The second Loomis depot (still standing) in 1915. A number of carts are loaded with boxes of fruit, such as peaches, cherries, and strawberries. The large boxes on the cart at the left are probably egg cases. The men are (left to right) express agent Earl Howard and depot agent Fred Martindale.

The Loomis Basin, located at the base of the Sierra Nevada foothills, was an early home to Indians, who lived in small villages along the creeks and streams—the Maidu tribe being the most dominant. History of the tribe and its way of life has been preserved in the Maidu Cultural Center in Roseville, California.

Trappers and hunters entered the area as early as 1825, a movement which began to change Indian life. New settlers brought fruit and nuts which were planted in the Bear River Valley. The rest is history.

The discovery of gold, of course, changed everything overnight but it was fruit farming that provided long term economic stability. Placer County was formed in 1851 further enhancing business opportunities. Roads for stagecoaches began to fall into place, providing additional amenities like roadhouses for the convenience of travelers. These roads, enhanced by the advent of the CPRR in 1864 and the transcontinental railroad in 1872, provided access to eastern United States markets for the high quality fruit and paved the way for the Lincoln Highway in 1913.

Loomis was not always "Loomis." Origi-

Packing Sheds

Fruit packing sheds of the Fruit Growers Association, on the Lincoln Highway in Loomis, California, 2001.

195

1915 Fire

A spilled oil stove at 3:00 A.M. engulfed the entire May residence and livery stable in Loomis on July 12, 1915. A slight wind carried the fire to the Loomis Hotel and then swept throughout the entire business section of town. Water diverted from a train and a wind change saved the fruit sheds.

nally the town was Pino but was only for residents who lived in Smithville, which was formerly known as Pine Grove. If this sounds confusing, it gets better. The Central Pacific Railroad, because of mail confusion with Reno, Nevada, called it Loomis in honor of James Oscar Loomis, postmaster. Now they had Loomis depot. The town was still Pino and the school district was Smithville. On May 28, 1890, Frederick Turner, postmaster, moved to have all things to do with the town be called Loomis. And so it was, and is.

Of the many awesome achievements of the Lincoln Highway, not the least was its taking the weary California travelers, fresh from the Sierra-Nevada challenge gently leading them into the beautiful, warm land, flowing with milk and honey—the Loomis Basin.

Cook's Store

Cook's store, built in 1912, at the corner of Taylor Road and Walnut Street. This building is now occupied by Main Drug Store. Note behind the store is the Loomis Livery Stable, all destroyed in the 1915 fire.

Loomis Basin Historical Society and the Leak Family, from the Harold & Kate Leak Collection.

OAKLAND, *California*

IN HIS OWN WORDS:

Friday, October 22, 1915

*Cloudy. Pulled into Oakland 42 miles.
Put car up at garage 745 E. 12th. At Oakland
and went over to Frisco on trolly.*

Sunday, October 24, 1915

Fair. Left for East. Stayed at Oakland. 3:30 P.M.

ENCINAL DEL TEMESCAL WAS THE NAME given to the Oakland region by early Spanish inhabitants. It meant, "Oak grove by the sweathouse," because of the luxurious growth of oak trees. When Oakland was chartered in 1852 the founding fathers liked the sentiment and chose the name "Oakland" and it stuck. In 1854 the town was incorporated. The strategic importance of Oakland was greatly accelerated in 1869 when it became the western terminus for the transcontinental railroad.

Oakland occupies an enviable spot within the rim of low hills along the eastern shore of the San Francisco Bay. Its average elevation is only 42 feet. Five thousand years prior to what we now know as Oakland, Indian children played along the shore of the bay and adults hunted along the creeks and in the hills. While these human dramas were being played out (maybe a bit over-romanticized), half a world away the Egyptian Pharoah, Menes, had founded the town of Memphis, but had not yet erected the pyramids. And in England, the bluestone/sandstone Stonehenge rocks would not appear for another 2,000 years. In perspective the Lincoln Highway with all its history and drama is hardly a drip in a chuckhole. This is not to treat those who gave us the Lincoln shabbily. The "Coast to Coast Rock Highway" made an invaluable contribution to us and our society within our own piece of time. However, who we are should not be without gratitude for those who have gone before.

The Ohlone or Costanoan peoples had jurisdiction of the shores of San Francisco Bay when the Spanish arrived in 1770 to explore the land and annex it to the Spanish Empire. It became a United States possession in 1848 along with the rest of California as part of the Treaty of Guadalupe Hidalgo.

Oakland grew rapidly in the twentieth Century. Following the disastrous San Francisco earthquake and fire in 1906, many families and businesses relocated east in Oakland across the bay. Modern Oakland is

Oakland Public Library, Oakland History Room
13th Street

Looking south on 13th Avenue toward the Inner Harbor from E.12th Street.

a great American city with a population of nearly 400,000 (up from 215,000 in 1914). Culturally diverse, it had all the excuses to strut: parks, museums, schools, business opportunities, recreation, arts, entertainment, transportation and a Mediterranean climate ranging in the 50s and 60s during winter months and to the70s and low 80s during the summer. Mickey Kantor, former United States Secretary of Commerce, said it well: "As the most diverse city in the nation, where more than 125 languages and dialects are spoken, Oakland is uniquely positioned as an excellent initiation point for international business seeking to develop trade relations in the U.S."

The prospect of the Lincoln Highway coming through Oakland created more than passing interest among local residents.

745 E. 12th Street, Oakland, where my father garaged his truck while he visited the Exposition. Original Lincoln Highway routes through Oakland differ, so this may or may not have been on the Lincoln in 1915. There is a good chance it was.

The following is one of many notices appearing in the *Oakland Tribune* on February 28, 1914:

Caine Consul for Lincoln Highway

Joseph E. Caine, secretary of the Oakland Commercial Club, was this morning officially named as California state consul for the Lincoln Highway, in charge of all highway work being done in the West. The appointment was received by wire from the East this morning, Caine being named in the place of E. P. Brinegar of San Francisco, who resigned two weeks ago. Following this comes the announcement that the official Western headquarters of the Lincoln Highway Association have been formally established at the Oakland Commercial Club, Oakland being deemed the terminal of the great road. The headquarters of H. E. Frederickson, consul at large for the United States, and Caine, as state consul, will be made in special offices on the seventh floor of the Hotel Oakland.

Oakland Public Library, Oakland History Room

Chevrolets, 1915

One day's output of Chevrolets at the Oakland, California, plant, 73rd. and Foothill, probably on the Lincoln Highway or at least within earshot of a backfire. Chevrolet was the first serious threat to Ford's Model T. Chevrolet became part of General Motors in 1918.

SAN FRANCISCO, *California*

IN HIS OWN WORDS

Saturday, October 23, 1915

Fair. Left early for Exhibition grounds.
Have covered everything but some state buildings.
At night seeing Frisco and China Town.
Some sight.

EVERY TOWN, PLACE OR EVENT has its heroes. Salt Lake City has its Brigham Young; Shelton, Nebraska—George Meisner; Evanston, Wyoming—James A. Evans; Columbia, Pennsylvania—John Wright; Lincoln Highway—Henry B. Joy and Carl Fisher. San Francisco has had many heros but let's pick one with some name recognition, a voice from the bay 200 years before San Francisco's first colonizing party in 1776. How about Sir Francis Drake ? Drake (1545-96) was the first Englishman to sail around the world (1577-1580) during which time he was a scourge to Spanish ships intent on invading England. His secret official mission was to plunder the gold-laden Spanish settlements and galleons and to challenge the Spanish monopoly of trade in the Pacific by establishing trade links with East Asia. Drake set sail from England with five ships (please note that they went west) sailing around the southern tip of South America. During an earlier voyage in 1572 he marched across the Isthmus of Panama where he got his first glimpse of the Pacific from the top of a tall tree. Since waiting 300 years for the Panama Canal was not an option, he chose the treacherous southern route, emerging into the Pacific with only one ship. Four fell victim to the wild waters. Now, with just one ship, Drake sailed north, pillaging Spanish settlements and ships. While off the California coast he was forced to shore for repairs to his ship. He dropped anchor in the vicinity to San Francisco (Drake's Bay, twenty-five miles North of San Francisco) and was greeted by the Miwok Indians, who took him and his men for gods and crowned Drake king. They stayed for five weeks before continuing their trip westward, crossing the Pacific and Indian oceans to the Atlantic Ocean by way of the southern tip of Africa. He arrived home in England to a hero's welcome in November of 1580 where Queen Elizabeth I knighted him, Sir Francis Drake—an amazing story! Consider this: all power was

Oakland Main Library

Ambulances

Two ambulances from the Exposition Emergency Hospital pose with staff within the colonnades of the Palace of Fine Arts.

Golden Gate Bridge

The Golden Gate Bridge as seen from Lincoln Park, San Francisco.

from the wind blowing into pieces of cloth stretched above the ship; no camera man; no communication with the local press; no airplane escorts, and no Rand McNally maps.

However, before we strike up the band for Sir Francis, let's listen to the tune of amateur British historian Bob Ward, whose research theorizes that Drake spent the summer of 1579 not in San Francisco Bay, but at Whale Cove on the Oregon coast where, in collaboration with Queen Elizabeth I, he falsified his records in an effort to thwart the Spanish in their search for the fabled Northwest Passage. Whatever the truth, at the very least here were some very courageous men with great nautical skill who helped pave the way for modern world trade—warts and all.

Joshua Norton (1819-1880) may not have been a San Francisco hero but it was not for lack of trying.

> At the pre-emptory request of a large majority of the citizens of these *United* States, I Joshua Norton, formerly of Algoa Bay, Cape of Good Hope, and now for the last nine years and ten months past of San Francisco, California, declare and proclaim myself the Emperor of these United States—September 17, 1859.

San Francisco outdid itself this time. It gave birth to the first and only Emperor of the United States – Joshua Norton, born in London, February 14, 1819. Norton arrived in San Francisco from South Africa in 1849 with $40,000 in his pocket, which he later lost in 1859 in an attempt to corner the San Francisco rice market. A colorful figure, this Norton. During the twenty-one years of his "reign," many decrees flowed from his royal pen. Some decrees were in jest by newspaper editors, but many were authentic. On July 16, 1860, he decreed the dissolution of the United States of America. On August 12, 1869, a Norton decree dissolved the Democratic and Republican parties because of party strife. On March 23, 1872, he ordered a suspension bridge be built between Oakland Point and Goat Island, and then on to San Francisco. If he had lived long enough, what he would have decreed about the Lincoln Highway is anyone's guess, but a decree of December 16, 1869, might give a hint. He decreed that Sacramento clean its muddy streets and place gaslights on the streets leading to the capitol. Emperor Norton I dropped dead on California Street at Grant Avenue on January 8, 1880. The funeral cortege was two miles long with upwards of 10,000 people.

In the nineteenth century San Francisco shared its passion for pelts with the rest of the West but with one significant addition: religious missions. The Franciscan Mission Delores dates to 1776 and is the oldest building in San Francisco—Registered Landmark Number One of the city. The first Mass was celebrated on June 29, 1776, just five days before the signing of the Declaration of Independence. The discovery of gold in 1848-49, of course, began a facelift for San Francisco and set the stage for the massive migration from east to west which created an increasing demand for good roads. The result was rails followed by roads morphing into the Lincoln Highway. All the above is pertinent. The rich history of Indians, the romantic name of Sir Francis Drake, the religious atmosphere of missions—all played an influential role in the choice of San Francisco, over New Orleans, Louisiana, as the site for the Panama celebration. Without the Panama Exhibition, what would Carl Fisher have used to fire the imagination of autoists and make a compelling case for the Lincoln Highway?

San Francisco was also host to what may have been the worst natural disaster in United States history (competing with the 1900 hurricane disaster in Galveston, Texas). The San Andreas Fault spoke up at 5:12 A.M. on April 18, 1906, followed by a devastating fire. The quake and conflagration resulted in 3,000 deaths and 225,000 injuries, with a monetary loss of $400,000,000 in 1906 dollars. Modern analysts estimate its intensity at 8.5 on the Richter scale. The ground broke open for more than 270 miles. Nine years later, in 1915, the city had not yet fully recovered, so the Panama-Pacific International Exposition was an important opportunity to restore the glory of San Francisco, which it did. The Exposition was a celebration of the completion of the Panama Canal and the 400th anniversary of the discovery of the Pacific Ocean by the Spanish explorer, Vasco Nuñez de Balboa

Lincoln Park

California Palace in Lincoln Park, the western terminus of the Lincoln Highway

(1475-1519) on September 13, 1513. He claimed the Pacific Ocean and all its shores for Spain.

The Exposition took over three years to construct, at a cost of $50,000,000, and ran from February 20 until December 14, 1915. A promotional brochure for Easterners sounded like it was written in Hollywood with its euphemistic descriptions and historical claims:

> This is the first time in the history of man the entire world is known and in intercommunication. In speaking of the earth, the qualification 'The *known world*' is no longer necessary. For the first time all the world is known…think of the pleasure you will derive from two or three weeks to a month vacationing in California. You need an outing, free from care and responsibility. Here is a chance to combine travel, pleasure, rest and instruction. To cross the great American desert, now a land of cultivated farms; to climb the Rockies and the Great Sierra Mountains and feast your eyes upon their rugged grandeur; to visit California and the Pacific slope and delight in their sun-clad beauty and blooming fertility; to look upon the vast stretches of the mighty ocean that sweeps to the shores of the fabled East; and then to visit the jewel city of the world's greatest Exposition—magnificent, beautiful, alluring—this is your opportunity if you will but take it.

And take it thousands did—adventurous souls suckered into braving the Lincoln Highway in 1915 by the exposition's glossy brochures, or Carl Fisher's hoopla, or just looking for an excuse to join the herds of lemmings going west. Not least among them was the "Paramount Girl", Anita King, a traveler one-of-a-kind. Her decision to cross the continent on the Lincoln Highway was not influenced by Centennial glitter or the magic of Carl Fisher, assuming that she even knew who he was. Instead she accepted the challenge of her boss, Jesse Lasky, motion picture magnate, who predicted it would be ten years before the Lincoln Highway would be improved enough for a lady to make the trip without difficulty. Anita's previous experience as a race car driver before trying her hand at the movie trade could not let such a left-handed slight on femininity go without an appeal to arms. She declared she would not only do it but do it all by herself. *The Los Angeles Times* explained it this way, "There will be nobody with her at any time on the trip. She will

Lincoln Highway National Museum & Archives

Anita King

Anita King makes history as she arrives in New York, October 19, 1915, in her Kissel Kar after her transcontinental drive over the Lincoln Highway from San Francisco

have no mechanician, no chauffeur, no maid. Her only companions will be a rifle and a six shooter."

Being a good sport, Lasky was equal to the occasion and proposed to pay for the trip and promised a high powered (31.6-hp) Kissel Kar as a prize if she pulled it off.

Anita King became not only the first woman to break the cross-country barrier by herself but she did it from west to east in Lasky's Kissel Kar, carrying a message from San Francisco Mayor Rolf, Jr., for New York Mayor John Mitchell. Her adventure began in San Francisco on Wednesday morning September 1, 1915, and ended forty-nine days later, on Tuesday October 19, 1915, in New York City. She traveled all twelve Lincoln Highway states, but not without some special moments. Four days out she hit Nevada and got stuck in mud following a sudden cloudburst. She shoveled mud from 9:00 A.M. until 8:00 P.M., got the car out, drove 50 feet and got stuck again. What happened next may be a close relative to green cheese on the moon. Exhausted and without food, she got some blankets to lie down when she was attacked by a coyote at midnight. After an incredible struggle she managed to shoot and kill the beast and then passed out. Some prospectors, hearing her shots, found her at 3:30 A.M., took her eighteen miles to a station house where she ate some food and rested before returning for her car and continuing on her way.

She was met in New York at the 129th Street ferry and escorted to City Hall to deliver the mayoral dispatches.

The "Paramount Girl" and my father probably passed each other somewhere near Grand Island, Nebraska.

Modern San Francisco is a major United States city with a population of more than 724,000. The Lincoln Highway is honored to be punctuated at each end with great American cities—New York and San Francisco. Concerning the latter, the most important spot, where it all comes together, is Lincoln Park. Unfortunately, my father didn't visit the park. The impression left in his diary was that half the fun was in getting there. In fact he only spent one day in San Fransisco—Saturday, October 23, 1915. In that one day he said he covered everything except some state buildings. He probably had not read the brochure. That night he went to Chinatown and the next day left Oakland for home, Sunday October 24.

Final Post

The final post is hiding in a bush at the corner of 132nd and El Camino in San Francisco.

FRANK C. NISSLEY

1882-1981

IF HE HAD HAD A MIND TO, MY FATHER could have chased his relatives back to 1594 in Switzerland (That was 200 years before the Philadelphia-Lancaster Turnpike). The Nissley (Nussli in Swiss-Deutch) family records, like so many others during that era, were destroyed by the social/religious/political upheaval during the sixteenth century European Reformation. My father was the fifth generation in the United States.

Frank Nissley was one of four boys and two sisters that grew up on a farm in Lancaster County, Pennsylvania. At age 21 in 1903 he left home to seek his fortune through the lens of a camera. In 1917, he hired widow Mary Jessica Madden, another photographer, who had been recently burned out in a hotel fire in Shamokin, Pennsylvania, and who was now looking for a job. That was a good move my father made, because in 1921 Mary Jessica became my mother. He and my mother made a good team snapping their shutters throughout eastern

Frank C. Nissley

Frank C. Nissley, business man, about 1905

Pennsylvania in the summer and Florida in the winter, but with the advent of a daughter and son, they decided the better part of wisdom was to settle back down on the farm—which they did, in 1924. My childhood memories were generated on the farm and later in Tampa, Florida, where we moved in 1932 as a consequence of the famous "29 crash." During the ensuing years we alternated between Florida and Pennsylvania, depending on the availability of work, finally settling in eastern Pennsylvania in the environment of the Lincoln Highway. My mother died in 1963 at age 81, and my father at age 99.

Frank C. Nissley

My father, Frank C. Nissley, stands on the doorstep of his ninetieth year. He died nine years later at age ninety-nine.

POSTSCRIPT

Westinghouse Bridge

This 1930s bridge connected not only Pittsburgh communities but also bridged the world of country cafes, roadside cabin courts, and gas stations to a post WW II age of Interstates and shopping centers.

Chester, West Virginia

The world's largest teapot. Photo taken during the 2004 Lincoln Highway conference in Chester.

Glenfield Brick Remnant

Looking west on the 1916 yellow brick road in Glenfield, Pennsylvania, northwest of Pittsburgh.

Chester, West Virginia

201 Virginia Avenue, in front of the Church of Christ. We could not find this post on our earlier visit because the church custodian had removed it for safe keeping during some church renovation.

Chester, West Virginia

On the west side of California Avenue, north of First street. This post was found by Dr Pugh on a trash heap somewhere in Pennsylvania. He inquired about its destiny and was told it would be thrown away. "That's a piece of history," said Pugh, and he was given the post. His son painted all three posts in town for the 2004 LH conference.

East Liverpool, Ohio

Fifth street and Broadway, in front of the Ceramic Museum, the former post office.

LINCOLN HIGHWAY *The Road My Father Traveled*

RESOURCES

PENNSYLVANIA
- US Route 30 Corridor Improvement Study
- Amish Country News
- Pennsylvania Department of Transportation
- *The Wayside Inns on the Lancaster Roadside Between Philadelphia and Lancaster,* by Julius F. Sachse (p. 82)

Does The Old Geezer Still Drive?
- *Setting Older Drivers Straight,* by Mark Zaloudex

Susquehanna River Bridges
- Columbia Historic Preservation Society

Columbia/Wrightsville, Pennsylvania
- Columbia/Wrightsville Preservation Society
- Hellam Museum, Historic Wrightsville Newsletter
- Susquehanna Valley Chamber of Commerce
- Kreutz Creek Preservation
- Columbia Civil War Centennial
- Historic Wrightsville Newsletter

Gettysburg, Pennsylvania
- Adams County Historical Society
- Battle of Gettysburg Center.

Chambersburg, Pennsylvania
- Buchanan's Birthplace, Commonwealth of Pennsylvania
- Chambersburg Chamber of Commerce
- Bureau of State Parks
- Kittochtinny Heritage Museum
- Chambersburg Free Library
- Franklin County Tourist Council

Everett, Pennsylvania
- Bloody Run Historical Society
- Chelsea Dunkle
- Jeffery W. Whetstone

Bedford, Pennsylvania
- Bedford Chamber of Commerce
- Pioneer Historical Society
- Bedford County Visitors Bureau
- Bedford County Library

WEST VIRGINIA
- *A History and Road Guide of the Lincoln Highway in Ohio,* by Michael Gene Buettner

Autocar Trucks
- *The Autocar History,* by Cheryl McPhilinny, McPhilinny Associates

OHIO
- *A History and Road Guide of the Lincoln Highway in Ohio,* by Michael Gene Buettner
- Standard Catalogue of American Cars

Lisbon, Ohio
- Lisbon Chamber of Commerce

Minerva, Ohio
- *Motoring, A Pictorial History of the First 150 years,* Peebles Press
- *Cars, Cars, Cars, Cars,* Paul Hamlyn—London
- Minerva Area Historical Society

Wooster, Ohio
- Wooster Chamber Of Commerce
- Wooster Community Guide, 2000
- Wayne County Historical Society

Packard
- The Packard Club
- The Packard Museum
- *Packard's 100th Anniversary,* by Terry Martin
- Motorsports Digest
- Standard Catalogue of American Cars

INDIANA
- Complete Official Road of the Lincoln Highway
- Lincoln Highway in Illinois and Indiana
- Cruise IN
- Indiana Historical Society
- Elwood Haynes Museum

Fort Wayne, Indiana
- Old City Hall Museum, Fort Wayne
- *Standard Catalogue of American Cars*

La Porte, Indiana
- La Porte County Historical Society
- The Hoosier State Beneath Us
- *Standard Catalogue of American Cars*

Pre-WW II US Auto Manufacturing
- *National Safety Council Accident Facts*
- *Standard Catalogue of American Cars*
- *A Century of Cars,* by Scott Oldham

ILLINOIS
- *Illinois History;* An annotated Bibliography
- Illinois Lincoln Highway
- *Standard Catalogue of American Cars*
- *The Roads That Built America,* by Dan McNichol

206

RESOURCES

Fulton, Illinois
- Fulton Chamber of Commerce
- *A History of Whiteside County*, by Wayne Bastian

Chicago Heights
- *Chicago Pneumatic; the First Hundred Years*
- Chicago Heights Signal; November 1913
- *Illinois History*; An annotated Bibliography
- Illinois Lincoln Highway

Geneva, Illinois
- History of Geneva, Illinois, 1998
- Geneva Chamber Of Commerce

Dirt Roads
- *People Who Live At the End of Dirt Roads*, by Lee Pitts

The Importance of Good Roads
- *National Safety Council Safety Facts*
- *Accident Facts*, by Skip Yazwinsky

IOWA
- Transportation In Iowa, Iowa Department of Transportation
- Davenport Public Library
- Iowa Chapter, Lincoln Highway
- *Lincoln Highway Through Iowa*

Cedar Rapids, Iowa
- *A House Full of strangers*, by Nova Dannels
- The History Center
- *Tall Corn and High Technology*, by Ernie Danck
- Linn County Historical Center

Belle Plaine, Iowa
- Beverly Winkie

Scranton, Iowa
- State Archeologist of Iowa
- *The Lincoln Highway of Iowa*, by Gregory Franzwa

Carroll, Iowa
- This Month In Iowa History
- *Word Pictures of Early Carroll County*, by James Kerwin
- Carroll Chamber Of Commerce
- *Over 200 Post Cards From Carroll*, by James Kerwin
- Carroll County Historical Museum
- Carroll Public Library

Arion/Dow City, Iowa
- Crawford County, GenWeb Project
- *125 years of Dow City-Arion History 1869-1994*, items extracted by Cindy Simon

Dunlap, Iowa
- Harrison County, Iowa History
- *The Loess Hills: A geologic View*, by Joan Cutler and Deborah J. Quade
- Loess Hills State Forest, http://www.state.iaus/government/dnrorganize/forest/ihsf.htm

Missouri Valley, Iowa
- Missouri Valley Area Centennial

NEBRASKA
- *Andreas' History of the State of Nebraska*, by Gary Martens and Laurie Saikin, http://www.ukans.edu/carrie/kancoll//andreas_neerlyhst-pl.html
- Wisconsin State History

Omaha, Nebraska
- Greater Omaha Chamber of Commerce

Valley, Nebraska
- Valley Community Historical Museum

Schuyler, Nebraska
- Schuyler, Nebraska Resource Guide
- Schuyler Museum
- Schuyler Chamber of Commerce

Shelton, Nebraska
- *Andreas' History of the State of Nebraska*, Buffalo County – part 9

Kearney, Nebraska
- *Andreas' History of the State of Nebraska*, Buffalo County – part 3
- History, Kearney, Nebraska, Nebraska Public Power District
- *Standard Catalogue of American Cars*

Elm Creek, Nebraska
- Nebraska State Historical Society
- Elm Creek Centennial Collection
- Frank Erickson Insurance Agency

Overton, Nebraska
- Overton Observer
- Nebraska State Historical Society
- Soul Of America-Black Towns: Nebraska

Paxton, Nebraska
- Nebraska Fence Post
- Fae Christiansen

Sidney, Nebraska
- Cheyenne County Historical Association

WYOMING
- *Trails to Tracks to Highways*
- *A Guide to the State of Wyoming—History*
- Wyoming Visitor Directory, 2001
- *Wyoming Tales and Trails*

Pine Bluffs, Wyoming
- *Pioneer Parade Vol. II*, by Martha Thompson
- The Handbook of Texas Online: Texas Revolution

Burns, Wyoming
- *Pioneer Parade Vol. II*, by Martha Thompson

Carbon, Wyoming
- *Ghost Towns of Wyoming*, by Donald C. Miller
- *Wyoming Tales and Trails*
- *The Lincoln Highway in Wyoming*, by Gregory Franzwa

Wamsutter, Wyoming
- *1890-1990 Wyoming Centennial, A Lasting Legacy*, by Janice Potts and Debra Dennis
- *1924 Official Road Guide of the Lincoln Highway*

Evanston, Wyoming
- City of Evanston, Wyoming
- Evanston Chamber Of Commerce
- *Wyoming Tales and Trails*
- *Lincoln Highway, Forum, Winter/Spring 2001*

Important Events, 1915
- The Panama-Pacific International Exposition, http//:www.sanfranciscomemories.com/ppie/1915.html
- Any Year In History, http//:www.scopesys.com/cgi/anyyear.cgi
- *Timetables of American History*

UTAH
- *Lincoln Highway, Utah*, by Gregory Franzwa
- Utah Division Of State History
- *U. S. "Theft" of Mexican Territory*, by Chris Scheffler
- *Wyoming Tales and Trails*
- *Roadside History of Utah*, by Cynthia Larson Bennett
- Grantsville Chamber of Commerce

Salt Lake City, Utah
- *Lincoln Highway History*—IL 02
- *Brief History and Description of Salt Lake City*
- *Brief History of Utah*, Utah Division of State History
- Salt Lake Tribune

NEVADA
- Cedar Rapids Gazette, July 3, 1983
- *And In The Beginning*, by James Shown

Eureka, Nevada
- Eureka County, Nevada—Official Home Page
- Economic Development Council, Eureka, Nevada
- Eureka County Museum
- Eureka Sentinel Museum

Austin, Nevada
- Austin Chamber Of Commerce
- *Lost In Austin*, by Jim Andersen
- Austin-Nevada Ghost Town

Fallon, Nevada
- Churchill Economic Development Authority
- Churchill County Museum and Archives
- Greater Fallon Chamber Of Commerce
- *Lincoln Highway Forum, VOL. 1, No. 4*

CALIFORNIA
- *History of the Donner Party*, Truckee Almanac
- The Donner Party, Brief History
- Elisha Stevens and the Donner Summit: Pioneers and Treacherous paths
- History of Truckee
- *Leon Schegg, Lincoln Highway Forum, Vol. 2, No. 2, 1995*
- World Almanac For Kids

Loomis, California
- Loomis Basin Chamber of Commerce
- Loomis Basin Historical Society
- *History of Placer County*, by Thompson and West

Oakland, California
- Oakland History.com
- Oakland Metropolitan Chamber of Commerce
- San Francisco Examiner
- Oakland Public Library

San Francisco, California
- San Francisco History
- San Francisco Chamber of Commerce
- Museum of the City of San Francisco
- *Famoous Hispanics*, by coloquino.com

Addendum

Tom Webber

To my knowledge, Tom has never experienced a face-to-face encounter with the Lincoln Highway, but he has faced computers head-on. Since my computer and I are often not on speaking terms, Tom has become a tireless and patient mediator and consultant, especially in the delicate enhancement of photographs.

Mildred Thompson

Long time author and editor, Millie provided invaluable professional editorial counsel and moral support. What have I done to merit the attention of one so able and yet so gracious? If there is any sparkle or subtlety of professionalism in this volume it's the Magic Touch of Millie Thompson.

LINCOLN HIGHWAY *The Road My Father Traveled*

Page numbers in **bold face** *refer to photographs.*

INDEX

1835 N. 20th. St., **111, 112**
21 miles to P, **9**
Abraham Lincoln, 26, **27**, 34, 115, 125, 140, 175
Adams, John Quincy, 24, 31
Advertisements, 2,4,5,10,35
Alkia Flat, **134**
American Auto Association, 3
American Revolution, 31
Ames Monument, **143**
Amish settlement, Wooster, 58
Anderson, B. L. (Babe), 174
Anderson, Waldemar, home of, **158**
Andrew Jackson, 70
Arion bank, **100**
Arion depot, **101**
Arion fires, 102
Arion post office, **99**
Arion, decline of, 102
Arion, Iowa, history of, 99, 100
Arion, Iowa, name of, 100
Arion/Dow city, Iowa, 99-102
Asp, Lynn, **73, 74**
Austin, Nevada, 179-182
Austin, Nevada, history of, 179
Author's Lincoln Highway roots, 8
Auto production in Ohio, 47
Auto registrations, Iowa, 83
Autocar, 46
Automotive Heritage Museum, 63
Autos 1915, 107

Baker, John, 79
Bank of Minerva, **56**
Banta, Jay, **167**
Barber, Robert, 17
Barndollar, Michael, 33-35
Battle of Gettysburg, 23, 24
Bedford Springs Hotel, 39,40
Bedford, Pennsylvania, 38-42
Belle Plaine, Iowa, 91, 92
Benjamin Harrison, 140
Bernard, Jane, **123, 124**

Bicycle ride from New York to San Francisco, 185, 186
Bloody Run Creek, 34
Bluston, Samuel, 17
Boot Hill, *112*
Boy falls from car on Lincoln Highway, 185
Brainnerd, Rev. Frank G., 170
Brick manufacturing in Ohio, 52
Brick, source of, 53
Bridges of Columbia /Wrightsville,16
Bryan, William Jennings, **150**
Buchanan and Shell Creek station, 119, 120
Bugas, Andrew P., 155
Burma Shave, 43
Burning bridge, **17**
Burns Wyoming, water tower, **149**
Burns, Wyoming 1915, **150**
Burns, Wyoming, **148**
Burns, Wyoming, 148-150
Butko, Brian, 1

CALIFORNIA, 189-203
Callao, 163, **166**
Car in the mud, **81**
Car off Lincoln Highway, **133**
Carbon bank, **153**
Carbon cemetery, **151, 152**
Carbon, Wyoming, 151-153
Carbon, Wyoming, history of, 151, 152
Carroll fire, 1879, 96, 97
Carroll, Iowa, 95-98
Carroll, Iowa, center of operation for my father, 97, 98
Carroll, Iowa, Lincoln Highway, **98**
Cedar Rapids, **87**
Cedar Rapids, history of, 87, 88
Cedar Rapids, Iowa, 87-90
Chambers, Benjamin, 31
Chambersburg jail, 30
Chambersburg, Pennsylvania, 30-32
Chapel, College of Wooster, **59**
Chapman, John, "Johnny Appleseed", 68
Chester, West Virginia, 44, 45
Chevrolet production 1915, **198**
Chicago Heights, Illinois, 75-76
Chicago Northwestern Railroad depot, 78

Chief Littleturtle, **68**
Chinatown, 156, 157, 203
Choice of name, Minerva, 55
Christiansen, Fae and Jean, Gene, **132**
Christmas 1928, 89, 90
Church Butte, **142**
Churches,
    Congregational, 77
    Disciples Of Christ, 77
    German Reformed, 31
    Lutheran, 29, 31, 147
    Methodist, 36, 37, 118
    Presbyterian, 31, 36, 58, 147
    Swedish Lutheran, 77
    Unitarian, 77
Civil War, 17, 23, 24, 26, 31, 106,
Coast-To-Coast Rock Highway, 3, 25, 65, 84, 197
Coffee pot, Bedford, **41**
Collins, Rev. A., 125
Columbia roundhouse, *22*
Columbia/Wrightsville, Pennsylvania, 17-23
Columbus, Christopher, 20
Commercial Club, 78
Concrete bridge, 19
Concrete marker posts, **4, 9, 11-15, 20, 29, 32, 36, 41, 46, 49-51, 56, 66-68, 73-74, 83, 85-86, 92, 94, 98, 103, 106, 109-111, 127, 135, 144, 155, 167, 177, 187, 193, 203**
Concrete surface, Iowa, 83
Confederate Forces, 23
Conoco station, **154**
Courthouse, Wooster, **59**
Cove Gap, 31
Cultural mix, settlers, 120

Dannels, Nova, 88-90
Death rate, auto accidents, 72
Dement House, 79, 80
Denny, John Q., 23
Devil's Den, **25**
Dirt Roads, 138
Does The Old Geezer Still Drive?, 15
Donna Reed cabin, **160**
Donner Lake, **189**

210

Page numbers in **bold face** *refer to photographs.*

Donner Party, 189
Donner Pass, 189, 191, 192
Dreamland Theater, **55**
Duesenburg, 68
Dunlap, Iowa, 103, 104
Dunlap, Iowa, early settlers, 103, 104
Dunlap, Iowa, history of, 103
Dutch farm lands, **6**

Earthquake 1906, 201
Eisenhower, Dwight & Mamie, 93
Elevators, **93**
Elm Creek theater, **129**
Elm Creek, Nebraska, **128-129**
Elm Creek, Nebraska, history of, 128
Espey House, 39
Eureka, Nevada court house, **177**
Eureka Sentinel, 178
Eureka, Nevada, 176-178
Eureka, Nevada, history of, 176, 177
Evanston, Wyoming, 156-158
Evanston, Wyoming, history of, 156, 157
Everett, Edward, 34
Everett, Pennsylvania, 33-37
Exposition ambulances, **199**

Fallon Nevada, Main Street, **184**
Fallon, Nevada, 183-187
Fallon, Nevada, history of, 183
Fell, Dr. Bessie, 148, 149
First license tag, **178**
First Packard, **60**
Fish Springs, 11, **164**
Fisher, Carl G., 3-4, 7, **64-66**, 84, 187, 199, 201-202
Flood, Valley, Nebraska, **117**, 118
Forbes road, **39**
Forbes, John, **39**
Ford garage, **120**, **129**
Ford, Henry, 4, 7, 33, 82
Forerunner of American Red Cross, **180**
Fort Bedford, **40**
Fort Benjamin Chambers, 30
Fort Pitt, 39
Fort Wayne, Indiana, 68-69
Founders Crossing, **40**

Four sentinels, **152**
Franciscan Mission Delores, 201
Franklin Grove, Illinois, 73
Franzwa, Gregory, 1, 4
Frenchman Station, **186**
Fruit sheds, 195
Fulton, Illinois, 79-80
Fulton-Lyons bridge, **79**

Gary North machine shop, 173, **174**
Geneva, Illinois, 77-78
Geneva, Illinois, cultural makeup, 78
Geneva, Illinois, history of, 77
George Preston station, **92**
George Washington, 8, 39, 68, 125
Gettys, James, 24
Gettys, Samuel, 24
Gettysburg Address, 26
Gettysburg, Pennsylvania, 24-29
Golden Gate Bridge, **200**
Goodyear Cutoff, 161
Granite Museum, **194**
Grantsville, Utah jail, **160**
Green's Hotel, **102**
Gridley Museum, **180**
Guinn, hyrcamus, and others, 91

Haunted House, **8**
Haynes Apperson, 63
Haynes, Elwood, **63**
Hearts of gold cantaloupes, 184
Heath, Mildred, **130**
Heider tractor, **97**
Hendricks, H. B., 1854, 105
Herrington, James, 77
Hoag, Gail, 4, **62**, 183
Hokanson, Drake, 1
Holdeman farm house, **8**
Hurricane bridge, **18**
Hybrid gas/electric, first, 71

ILLINOIS, 73-80
Illinois, history of, 73
Importance of good roads, 81
INDIANA, 63-71
Indiana Ideal section, **66**, 73-80

Indianapolis speedway, 65, **66**
Indianapolis, history of, 68
Indians of Columbia/Wrightsville, Onandago, Seneca, Oneida, Tuscarora, 23
Indians of Gettysburg, Shawnee, 17, Iraquois, 24
Indians, 70, 68, 73, 82, 87, 99, 103, 115, 136, 145, 183, 190, 197, 199
International Hotel, **182**
IOOF building, **121**
IOWA, 82-107
Iowa, history of, 82
Iowa mud, **136**

James Buchanan, 31, 39, 175
Jamison, Mary, 30
Jefferson tower, **83**
Joy, Henry B., 1, 3-5, 7, 60-62, 65, 111, 199
Joy, Henry, and Bement, Austin, **111**
Joyner, Pitts, **192**
Joys of motoring, **82**
Judiciary Act, 31
Juniata River, **38**

Karns Garage, **33**
Karns, A. W., 33
Karns, W. C., 33
Kearney, Nebraska, 125-127
Kearney, Nebraska, history of, 125, 126
Kearney's Ranch, **166**
Keeley Stove truck, **21**
Kent, I. H., 183
Kimes, Beverly Rae, 72
King, Anita, **202**, 203
Kinney Lewis, 52
Kissel Kar, **202**, 203
Kittochtinny Historical Society, 30
Kutz, Kevin, 40

La Porte, Indiana, 70-71
Lancaster County, Pennsylvania, 23
Lancaster, provisional capital, 7
Landseekers, **137**
Lee, Robert E., 31

211

Page numbers in **bold face** *refer to photographs.*

Lepper Library, **52**
Lincoln Café, **92**
Lincoln Corridor Parkway, 9
Lincoln Highway Association national office, 73
Lincoln Highway Association, 3, 73, 74, 76
Lincoln Highway Association, revival of, 4
Lincoln Highway garage, **193**
Lincoln Highway history in Indiana, 63
Lincoln Highway in the desert, **162, 163**
Lincoln Highway renovation, **92**
Lincoln Highway repair, **89**
Lincoln Highway through Utah, 160, 161
Lincoln Highway towns in Illinois with auto production, 73
Lincoln Highway towns in Indiana, 63
Lincoln Highway traffic, **90**
Lincoln Highway west of Carbon, **152**
Lincoln Highway, Everett, Pennsylvania, **37**
Lincoln Highway, Scranton, Iowa, **94**
Lincoln Highway, the second continental Llink, 116
Lincoln Monument, **140**
Lincoln Park, **201**-203
Lincoln Trail, 179, 180
Lisbon, Ohio, 52-53
Little Giant truck, 1, 3, 24, **25**, 73, **75**, **76**, 128, 130, 132, **143**, 181, **185**
Little Giant truck, history of, 75, 76
Lockhart, John, 23
Loneliest Road, **175**
Longest road, 80
Loomis California, name, 196
Loomis, California, 194-196
Loomis, California, fire, 1915, **196**
Loomis, California, history of, 194-196
Lost Gold Village, 57
Low fuel, 172
Lyric theater, **22**, 23

MaAdam, 7, 83
Mac Ray, Robert, 38, 39
Macadam surface, 83
Marsh Creek settlers, 24
Martin Van Buren, 70
Maxen electric, **88**

Maybach, Wilhelm, 56
Mc Daniel building, **34**
McCausland, Gen. John, 31
Meisner bank, **122**, **123**
Mercedes, history of, 56
Meyer farm, **89**
Mike Fallon Ranch, 183
Miller, John, 23
Minerva, history of, 54
Minerva, Ohio, 54-57
Missouri Valley, Iowa, 105, 106
Missouri Valley, Iowa, history of, 105, 106
Missouri Valley, Iowa, pioneer businesses, 105
Modern Lisbon, Ohio, 52
Mormon Crickets, 164
Mormons, 115, 122, 160, 162, 168, 169, 171, 199
Mother Bedford, 39
Mountain Pass, **161**
Munson factory, **70**
Munson Omnibus, **70**
Munson, John W., 70, 71
Musser, John, 33
Mutual aid on the Lincoln Highway, **137**
My Auto Without Thee, 188
My father's diary, 1, 17, 21, 24, 30, 33, 38, 52, 54, 58, 68, 70, 75, 77, 79, 87, 91, 93, 95, 99, 103, 105, 108, 114, 115, 117, 119, 122, 125, 128, 130, 132, 136, 145, 148, 151, 154, 156, 168, 176, 179, 183, 194, 197, 199

Named highways, 4,
Nation's capital, 20
NEBRASKA, 108-135
Nebraska, History of, 108
Nebraska, people of, 109
NEVADA, 172-187
Nissley, Frank C., 1, 3, 17, 21-25, 28, 30, 32, 33-35, 37, 38, 40, 44, 47, 48, 52-54, 58, 59, 68, 70, 73, 75, 77-79, 87, 91, 93, 95, 97-103, 105, 108, 112, 114-119, 122, 125, 126, 128-132, 136, 139, 143, 145, 147, 148, 151, 152, 154, 156, 162, 163, 165, 168, 174, 176, 179, 181, 183-186, 191, 194, 195, 197-199, 203, 204

Norton, Joshua, 200, 201
Numbered highways, 4, 47
Oakland California, history of, 197
Oakland California, Thirteenth Street, **197**
Oakland, California, 197, 198
Oakland, California, Twelfth Street, **198**
Ohio River bridge, 45
OHIO, 47-59
Oil discovery, 58, 59
Old gas station, Minerva, **55**
Old stone House Museum, **53**
Old stone tavern, 52
Omaha, history of, 116
Omaha, Nebraska, 114-116
Orpheum theater, 32
Orr's Ranch, **161**
Osterman, H. C., 180, 186, 187
Osterman, Henry, Memorial, **65**
Overland Hotel, **184**
Overton bridge, **131**
Overton Herald, **131**
Overton, Nebraska, 130-131
Overton, Nebraska, history of, 130
Owens, Bob, **84**

Packard firsts, **62**
Packard, 5, 60-62
Palisades, **143**
Panama-Pacific International Exposition, 1, 3, 25, 65, 76, 201
Pardington, A. R., 76, 187
Pastime theater, **147**
Patocka, Shirley, **194**
Paxton garage, **133**
Paxton, Nebraska, 132-135
Paxton, Nebraska, name of, 132
Penn, William, 23, 24
PENNSYLVANIA, 7- 42
Philadelphia to Lancaster turnpike, 9
Photo, $75.00, **113**
Pillars of Ohio, **48**
Pine Bluffs, Wyoming, 1915, **145-147**
Pine Bluffs, Wyoming, history of, **145-147**
Pioneer hardships, 52
Pioneer Historical Society, 40

Page numbers in **bold face** *refer to photographs.*

Pioneer life, 55
Platte and Elkhorn rivers, 117,118
Population of Lisbon, 52
Prairie schooner body, 116
President, William Waddington, 76
Pre-WW II Auto manufacture, 72

Quaker, 17, 54

Railroad bridge, **18**
Railroad car manufacturing, 56
Railroad construction, **183**
Railroad tower, **101**
Rain and drought cycle, 133
Rainbow bridge, **192**
Red Rose churches:
    German Reformed, Lutheran,
    Presbyterian, 31
Rehwinkel, Rev., Alfred, 148, 149
Remnants, 7, **45**, **47**, **58**, 109
Ribbon cutting, Elkhorn, Nebraska, **112**
Rich, Jacob, 23
Riggles, Joe, 40
Ristine, Burton, 1917 Hudson, **112**
Road to Carbon, **152**
Rock 'N' Rye, **158**
Rock Springs Pass, **172**, **173**
Rock Springs road, **172**
Ronald Reagan, 80
Ross, Nellie Tayloe, 141

Salt Lake City, Main street, **168**
Salt Lake City, Utah, 168-171
Salt Lake City, Utah, history of, 168, 169
Salt Lake, **170**
San Francisco, California, 199-203
Sargent, Merrill G., **146**
Saxon, **3**, 10
Schuyler Sun, 121
Schuyler, name of, 119
Schuyler, Nebraska, 119-121
Schuyler, Nebraska, city hall, **120**
Schuyler, Vine Street, 119
Scranton, Iowa, 93, 94
Seedling mile, first, 73
Seedling miles, 9, 73, 84

Seiberling, F. A., 3
Seseney, Dr. Benjamin, 31
Seven-mile stretch, **42**
Shawanatown, 20
Shelton, Nebraska, 122-124
Shepherd, Osgood, 87
Shepherd's Tavern, 87
Ship Hotel, **42**
Sidney, Nebraska, 136-137
Sidney, Nebraska, 1915, **136**
Sidney, Nebraska, bad reputation of, 136
Silk manufacturing, 20
Simpson Springs, **162**
Sintz engine, 63
Sir Francis Drake, 199
Snowsheds and lake, **190**
South Platte bridge, **134**
Spaulding, Payson, **156**
State Capitol, **169**
Steam engine, **98**
Steam plow, **150**
Stephen, Murphy, Townsend Party, 189-191
Studebaker, 63, 64
Supreme Court, 31
Susquehanna River, 17-23, **20**

Taggart, Stan, **158**
Tama bridge, **84**
Taylor, Pim, 54
Tea Pot, **44**
Texas cattlemen, 120
Texas longhorns, 145, 146
The Arch, **127**
The auto and Burns history, 148
The Great Race, **141**
The longest road, 80
Theodore Roosevelt, 183
Thomas Jefferson, 20, 52, 68, 115,175
Thomas Mifflin, 7
Thomas Telford, 7
Thomas, John, 165
Three-Mile-Picture-Show, 76, 180, 185-187
Top five states, auto production, 72
Treatment of Indians, 70
Treaty of Guadelupe Hildago, 160, 193, 197
Tree-In-Rock, **139**

Treinen garage, **137** Truck repair, 116
Ulysses S. Grant, 119
Underground railroad, 22, 30
Union Forces, 23
UTAH, 160-171
Utah, history of, 160

Valley, Nebraska, 113-114
Vasco Nunez de Balboa, 202
Vasques de Coronado, Francisco, 108
Victory, 1, 95, 125, 136, 145
Vogler, Hank, 173
Vogler, Pyrenees sheep dog, 173

Waldner, Ken, 155
Walters theater, 24, **28**, **29**
Wamsutter, Wyoming, 154-155
Wamsutter, Wyoming, history of, 154, 155
Wannamaker, John, 30
Ward, Bob, historian, 200
Washington Boro, 17
Washington D. C., 20
Wayne, Anthony, 34
Wendover Cutoff, 161
WEST VIRGINIA, 44-45
Western Pacific Railroad, 77, 78
Whitaker, John, 54
Whitaker, Keziah, 55
White Squaw marker, 30
Williams, Kay, 40
Winkie, Bev. and Wallace, 91
Wirtz Building, **53**
Wooster, Ohio, 58 – 59
Wright, John, 17-20
Wright, Susannah, 20
Wright's ferry, 20, 21
WYOMING, 137-156
Wyoming, history of, 139-141

Yoder, Jacob, 58
York County, Pennsylvania, 23
You Auto Know, 159
Young, Brigham, 171

213